STUDENT
ORGANIZER

JANA BRYANT
Daviess County High School

JULIE FRANCAVILLA
State College of Florida Manatee-Sarasota

DEVELOPMENTAL MATHEMATICS
SECOND EDITION

Elayn Martin-Gay
University of New Orleans

Prentice Hall
is an imprint of

Reproduced by Pearson Prentice Hall from electronic files supplied by the author.

1 2 3 4 5 6 BRR 15 14 13 12 11

ISBN-13: 978-0-321-64649-1
ISBN-10: 0-321-64649-5

Prentice Hall
is an imprint of

www.pearsonhighered.com

Table of Contents

Organizer Overview (for Instructors)

Greetings and thank you so much for using my Organizer. Upon completion of using this tool, I welcome any comments that will help me do a better job.

Let me describe to you what I see as a major problem in mathematics and the solution I am trying to provide:

Problem: Many of our students come to mathematics courses lacking not only the understanding of mathematical concepts and skills needed for high school mathematics, but also the general organizational/study tools needed for success. These helpful tools-for-success include general note-taking in mathematics, notebook organization, and basic study methods that are particularly successful in mathematics. You have probably noticed that many of our students, even those who participate successfully in the classroom, have trouble starting and documenting their assigned homework, reading their text, and maybe most importantly, keeping organized coursework.

Solution: This supplement is to be actively used with Martin-Gay's *Developmental Mathematics, 2nd ed.* If used properly, it will enable each student to grow in the skills listed above and to be successful and fully organized in their current course.

The Student Organizer contains:

> Organizer Overview for Students (printed front and back)

> For each section of the text, there are two 3-hole punched pages (printed front and back), with each page divided into sections focusing on needed organization.

> The final page for each chapter contains help for preparing for a chapter test.

> **Each student does need a three-ring binder (notebook), preferably with pockets.**

How does the Organizer Work?

Please read the Organizer Overview (for Students) for detailed instructions. You will see how this Organizer helps a student start reading this text, and start using some available supplements.

As an instructor, you can easily customize this supplement. Your course syllabus, etc., can be placed at the front of this notebook. If you give quizzes, students can place each with an appropriate section or in a notebook section solely for that purpose. Tests can be placed in the notebook pockets or in another designated section of the notebook.

Thank you, again, and best wishes for your Developmental Mathematics course,
Elayn (Martin-Gay)

Organizer Overview (for Students)

Greetings and thank you for using my Organizer.

This supplement is to be placed in a 3-ring notebook (preferably with pockets) to be used along with Martin-Gay's *Developmental Mathematics, 2nd ed.* If used as directed, this Organizer will help you become better organized, use your text more efficiently, use the Lecture Videos that are provided for you, and ultimately, increase your study skills for not only this course, but for other classes you are taking now and in the future.

How to Use this Supplement:

First, make sure you have the proper supplies.

Tools needed: Text, Video Lecture Series, writing instrument (pencil or pen), and this Organizer in a three-holed notebook.

For each section of each chapter, there are two 3-hole punched pages printed front and back, with each page divided into sections focusing on needed organization. These sections are structured as noted below.

Page 1 (front and back) contains:

> **Before Class**: Read and follow the given directions. Place a checkmark within the small open square before each set of directions as you complete each set.

> **During Class:** This section has to do with writing notes that will be helpful to you later as you start homework or study for a quiz or test. Notice that the remainder of this page along with the back of the page is divided into two columns.

>> **Class Notes/Examples**: In this column, write down any examples (line-by-line) demonstrated by your instructor, seen as an example in MyMathLab, or in the Lecture Videos.

>> **Your Notes**: This smaller column to the right is for your personal notes; for example, for you to write down things you don't want to forget.

> **After this page**, please insert any additional paper of your own to write any further notes.

Organizer Overview (for Students)

Page 2 (front and back) contains:

Practice: Read and follow the given directions. There are numbered examples and exercises for you to read and/or complete. For each of these, the answers and/or references are at the end of the section for your review. The types of exercises/examples are:

Review this example: This example is shown worked and completed and the answer is circled. Read this example and make sure you understand the solution.

Your turn: This exercise is for you to work and circle when completed. Make sure you check the answer at the end of this section. If correct, move to the next example/exercise. If incorrect, use the reference by the answer to view detailed steps of the solution.

Complete this example: This example is partially complete. Read the completed part and fill in the blank(s). (Follow the same steps as above to check and correct your work.)

After this page, please insert your paper containing your *written homework* * assigned by your instructor.

---

* *Written Homework*: Attempt all exercises asked of you. All odd answers are in the e-book, so make sure you check the answers to these exercises. If an exercise answer is incorrect, try to correct it on your own. If you are unable to correct an exercise, place a mark by the exercise number (such as a question mark, "?") so that you will know to ask your instructor about it. If there is an exercise that you want to make sure you study again before a test, place a mark (maybe an "!" mark) that you will recognize later.

Follow these directions as closely as possible. I know this may be difficult in the beginning, but trust that this Organizer can help you with your mathematics course.

Best wishes to you in your Developmental Mathematics course,
Elayn Martin-Gay

Section 1.2 Place Value, Names for Numbers, and Reading Tables

Before Class:

☐ Read the objectives on page 8.

☐ Read the **Helpful Hint** boxes on pages 9 and 10.

☐ Complete the exercises:

1. The number 932 is read as nine hundred _____ .

2. The number 9320 is read as nine _____ , three hundred twenty.

3. The number 9,320,000 is read as nine _____ , three hundred

   twenty _____ .

4. Add: $50,000 + 4,000 + 300 + 20 + 1 =$ _____

During Class:

☐ **Write your class notes.** Neatly write down **all** examples shown as well as key terms or phrases with definitions. If not applicable or if you were absent, watch the Lecture Series (DVD) for this section and do the same (write down the examples shown as well as key terms or phrases). Insert more paper as needed.

Class Notes/Examples	**Your Notes**

Answers: **1)** thirty-two **2)** thousand **3)** million, thousand **4)** 54,321

Section 1.2 Place Value, Names for Numbers, and Reading Tables

Class Notes (continued)	**Your Notes**

(Insert additional paper as needed.)

Section 1.2 Place Value, Names for Numbers, and Reading Tables

Practice:

☐   Complete the Vocabulary and Readiness Check on page 12.

☐   Next, complete any incomplete exercises below. Check and correct your work using the answers and references at the end of this section.

Review this example:	**Your turn:**
1. Determine the place value of the digit 3 in the number 93,192.	**2.** Determine the place value of the digit 5 in the number 657.

93,192
↑
The period is thousands.

The place value is ⟨thousands.⟩

Complete this example:	**Your turn:**
3. Determine the place value of the digit 3 in the number 534,275,866.	**4.** Determine the place value of the digit 5 in the number 5423.

534,275,866
↑
The period is millions.

The place value is ⟨_____.⟩

Review this example:	**Your turn:**
5. Write the whole number 27,034 in words.	**6.** Write the whole number 26,990 in words.

$$\boxed{27}, \boxed{034}$$

↓   ↓

_____ , _____
thousands   ones

⟨twenty-seven thousand, thirty-four⟩

3

Section 1.2 Place Value, Names for Numbers, and Reading Tables

Review this example:	**Your turn:**
7. Write nine thousand, three hundred eighty six in standard form.	**8.** Write fifty-nine thousand, eight hundred in standard form.

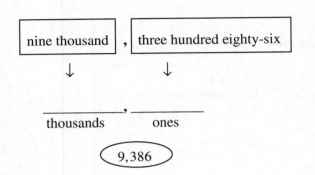

Review this example:	**Your turn:**
9. Write $2,706,449$ in expanded form.	**10.** Write $80,774$ in expanded form.

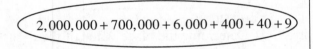

	Answer	Text Ref	Video Ref		Answer	Text Ref	Video Ref
1	thousands	Ex 2, p. 8		6	twenty-six thousand, nine hundred ninety		Sec 1.2, 3/6
2	tens		Sec 1.2, 1/6	7	9,386	Ex 10, p. 10	
3	ten millions	Ex 3, p. 8		8	59,800		Sec 1.2, 4/6
4	thousands		Sec 1.2, 2/6	9	$2,000,000 + 700,000 + 6,000 + 400 + 40 + 9$	Ex 12, p. 10	
5	twenty-seven thousand, thirty four	Ex 6, p. 9		10	$80,000 + 700 + 70 + 4$		Sec 1.2, 5/6

☐ **Next, insert your homework.** Make sure you attempt all exercises asked of you and show all work, as in the exercises above. Check your answers if possible. Clearly mark any exercises you were unable to correctly complete so that you may ask questions later. DO NOT ERASE YOUR INCORRECT WORK. THIS IS HOW WE UNDERSTAND AND EXPLAIN TO YOU YOUR ERRORS.

Section 1.3 Adding Whole Numbers and Perimeter

Before Class:

☐   Read the objectives on page 16.

☐   Read the **Commutative Property of Addition** box on page 17.

☐   Read the **Associative Property of Addition** box on page 18.

☐   Read the **Key Words or Phrases of Addition** chart on page 19.

☐   Complete the exercises: Fill in the chart.

	Key Words or Phrases	Examples	Symbols
1.	plus	9 plus 14	
2.	increased by	13 increased by 7	
3.	more than	5 more than 6	
4.	sum	the sum of 11 and 12	
5.	total	the total of 15 and 7	

During Class:

☐   **Write your class notes.** Neatly write down **all** examples shown as well as key terms or phrases with definitions. If not applicable or if you were absent, watch the Lecture Series (DVD) for this section and do the same (write down the examples shown as well as key terms or phrases). Insert more paper as needed.

Class Notes/Examples	**Your Notes**

Answers: **1)** $9+14$ **2)** $13+7$ **3)** $6+5$ **4)** $11+12$ **5)** $15+7$

5

Section 1.3 Adding Whole Numbers and Perimeter

Class Notes (continued)	Your Notes

(Insert additional paper as needed.)

Section 1.3 Adding Whole Numbers and Perimeter

Practice:

☐   Complete the Vocabulary and Readiness Check on page 22.

☐   Next, complete any incomplete exercises below. Check and correct your work using the
answers and references at the end of this section.

Review this example:

1. Add: $23 + 136$

Line up numbers vertically so that the
place values correspond.

$$
\begin{array}{r}
2\,3 \\
+1\,3\,6 \\
\hline
159
\end{array}
$$

Add digits in the ones place.
3 ones + 6 ones = 9 ones

Add digits in the tens place.
2 tens + 3 tens = 5 tens

Your turn:

2. Add:

a)
$$
\begin{array}{r}
5\,2\,6\,7 \\
+\ 1\,3\,2 \\
\hline
\end{array}
$$

b)
$$
\begin{array}{r}
8 \\
9 \\
2 \\
5 \\
+1 \\
\hline
\end{array}
$$

Look for
pairs of
numbers that
sum to 10.

Review this example:

3. Add: $1647 + 246 + 32 + 85$

Line up numbers vertically so that the place
values correspond.

When the sum of digits in corresponding place
values is more than 9, carrying is necessary.

$$
\begin{array}{r}
{\scriptstyle 1\,2\,2} \\
1\,6\,4\,7 \\
2\,4\,6 \\
3\,2 \\
+\ 8\,5 \\
\hline
2010
\end{array}
$$

Add digits in the ones place.
(7 + 6 + 2 + 5) ones = 20 ones
or 2 tens

Add digits in the tens place.
(2 + 4 + 4 + 3+ 8) tens = 21 tens or
2 hundreds + 1 ten

Add digits in the hundreds place.
(2 + 6 + 2) hundreds = 10 hundreds
or 1 thousand

Add digits in the thousands place.
(1 + 1) thousands = 2 thousands

Your turn:

4. Add: $24 + 9006 + 489 + 2407$

7

Section 1.3 Adding Whole Numbers and Perimeter

Review this example:

5. Before 2009, a stadium in Austin, Texas, could seat 94,113 fans. Recently, the capacity of the stadium was increased by 4525 seats. Find the new capacity of the stadium for the 2009 season.

The key word here is increased which means to add.

$$9\,4,1\,1\,3$$
$$+\ 4,5\,2\,5$$
$$\overline{9\,8,6\,3\,8}$$

Your turn:

6. Find the sum of 297 and 1796.

Complete this example:

7. Find the perimeter of the figure.

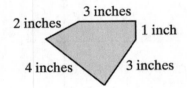

To find the perimeter (distance around), we add the lengths of the sides.

$P = 2\,\text{in.} + 3\,\text{in.} + 1\,\text{in.} + 3\,\text{in.} + 4\,\text{in.} =$

Your turn:

8. Find the perimeter of the figure.

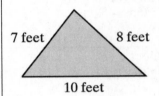

	Answer	Text Ref	Video Ref		Answer	Text Ref	Video Ref
1	159	Ex 1, p. 16		5	98,638 seats	Ex 7, p. 20	
2	a. 5399 b. 25		Sec 1.3, 1-2/6	6	2093		Sec 1.3, 5/6
3	2010	Ex 4, p. 18		7	13 in.	Ex 5, p. 19	
4	11,926		Sec 1.3, 3/6	8	25 ft		Sec 1.3, 4/6

☐   **Next, insert your homework.** Make sure you attempt all exercises asked of you and show all work, as in the exercises above. Check your answers if possible. Clearly mark any exercises you were unable to correctly complete so that you may ask questions later. DO NOT ERASE YOUR INCORRECT WORK. THIS IS HOW WE UNDERSTAND AND EXPLAIN TO YOU YOUR ERRORS.

Section 1.4 Subtracting Whole Numbers

Before Class:

☐   Read the objectives on page 28.

☐   Read the **Helpful Hint** boxes on pages 31 and 32.

☐   Read the **Key Words or Phrases of Subtraction** chart on page 31.

☐   Complete the exercises: Fill in the chart.

	Key Words or Phrases	Examples	Symbols
1.	difference	the difference of 12 and 4	
2.	decreased by	13 decreased by 7	
3.	less	6 less 5	
4.	less than	5 less than 6	
5.	subtracted from	8 subtracted from 10	

During Class:

☐   **Write your class notes.** Neatly write down **all** examples shown as well as key terms or phrases with definitions. If not applicable or if you were absent, watch the Lecture Series (DVD) for this section and do the same (write down the examples shown as well as key terms or phrases). Insert more paper as needed.

Class Notes/Examples	**Your Notes**

Answers: **1)** $12 - 4$ **2)** $13 - 7$ **3)** $6 - 5$ **4)** $6 - 5$ **5)** $10 - 8$

Section 1.4 Subtracting Whole Numbers

Class Notes (continued)	**Your Notes**

(Insert additional paper as needed.)

Section 1.4 Subtracting Whole Numbers

Practice:

☐   Complete the Vocabulary and Readiness Check on page 33.

☐   Next, complete any incomplete exercises below. Check and correct your work using the
answers and references at the end of this section.

Review this example:

1. Subtract and check by adding: $543 - 29$.

Line up numbers vertically so that the
place values correspond.

> In the ones place, 9 is bigger than 3, so
> we borrow from the tens place.

$$\begin{array}{r} \overset{3\ 13}{5\cancel{4}\cancel{3}} \\ -29 \\ \hline 514 \end{array}$$

Solution:

Check:
$$\begin{array}{r} 514 \\ +29 \\ \hline 543 \end{array}$$

$543 - 29 = \boxed{514}$

Your turn:

2. Subtract and check by adding:

a.
$$\begin{array}{r} 62 \\ -37 \\ \hline \end{array}$$

b. Subtract 5 from 9.

Complete this example:

3. The radius of Jupiter is 43,441 miles. The
radius of Saturn is 7257 miles less than the
radius of Jupiter. Find the radius of Saturn.

In Words:

> radius of Jupiter
> -7257
> ———————————
> radius of Saturn

> Replace the
> radius of
> Jupiter with
> 43,441 miles.

Subtract:

$$\begin{array}{r} 43,441 \\ -\ 7257 \\ \hline \end{array}$$

> The radius of Saturn is
> _____ miles.

Your turn:

4. The Oroville Dam is the tallest dam
in the United States at 754 feet.
The Hoover Dam is 726 feet high.
How much taller is the Oroville
Dam than the Hoover Dam?

Section 1.4 Subtracting Whole Numbers

Complete this example:

5. Subtract: $900 - 174$

In the ones place, 4 is larger than 0, so we borrow from the tens place. Since a 0 is in the tens place we will need to borrow from the hundreds place.

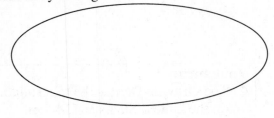

Borrow 1 hundred from the 9 hundreds. This leaves 8 hundreds.   1 hundred is the same as 10 tens. Add the 10 tens to the tens place.

Now borrow from the tens place.

Borrow 1 ten from the 10 tens. This leaves 9 tens. 1 ten is the same as 10 ones. Add the 10 ones to the ones place. Now, you are ready to subtract.

Check by adding:

Your turn:

6. Subtract and check by adding:

$$\begin{array}{r} 5\,1,1\,1\,1 \\ -\ 1\,9,8\,9\,8 \\ \hline \end{array}$$

	Answer	Text Ref	Video Ref		Answer	Text Ref	Video Ref
1	514	Ex 3, p. 30		**4**	28 *ft*		Sec 1.4, 6/6
2	a. 25 b. 4		Sec 1.4, 2&4/6	**5**	$726 + 174 = 900$	Ex 4, p. 30	
3	36,184 *miles*	Ex 5, p. 31		**6**	31,213		Sec 1.4, 3/6

☐ **Next, insert your homework.** Make sure you attempt all exercises asked of you and show all work, as in the exercises above. Check your answers if possible. Clearly mark any exercises you were unable to correctly complete so that you may ask questions later. DO NOT ERASE YOUR INCORRECT WORK. THIS IS HOW WE UNDERSTAND AND EXPLAIN TO YOU YOUR ERRORS.

Section 1.5 Rounding and Estimating

Before Class:

☐   Read the objectives on page 39.

☐   Read the **Helpful Hint** box on page 45.

☐   Read the **Rounding Whole Numbers to a Given Place Value** box on page 39.

☐   Complete the exercises:

1.   Circle the numbers that are 5 or greater.

    4   8   2   0   5   9

2.   In the number 1,023,475 what digit is in the tens place? _____

3.   In the number 1,023,475 what digit is in the thousands place? _____

4.   In the number 1,023,475 what digit is in the millions place? _____

During Class:

☐   **Write your class notes.** Neatly write down **all** examples shown as well as key terms or phrases with definitions. If not applicable or if you were absent, watch the Lecture Series (DVD) for this section and do the same (write down the examples shown as well as key terms or phrases). Insert more paper as needed.

Class Notes/Examples	**Your Notes**

Answers: **1)** 8, 5, and 9   **2)** 7   **3)** 3   **4)** 1

Section 1.5 Rounding and Estimating

Class Notes (continued)	Your Notes

(Insert additional paper as needed.)

Section 1.5 Rounding and Estimating

Practice:

☐   Complete the Vocabulary and Readiness Check on page 43.

☐   Next, complete any incomplete exercises below. Check and correct your work using the
        answers and references at the end of this section.

Review this example:	**Your turn:**
1. Round 568 to the nearest ten.	**2.** Round 635 to the nearest ten.

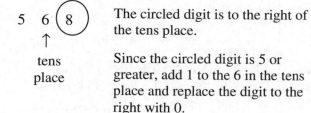

5   6   ⑧   The circled digit is to the right of
      ↑   the tens place.

    tens   Since the circled digit is 5 or
    place   greater, add 1 to the 6 in the tens
                  place and replace the digit to the
                  right with 0.

568 rounded to the nearest ten is ⟨570⟩.

Complete this example: **Your turn:**

3. Round 278,362 to the nearest thousand.   **4.** Round 36,499 to the nearest thousand.

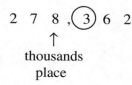

2   7   8 , ③ 6   2   Since the circled digit is
        ↑   NOT 5 or greater, do not
  thousands   add 1. Replace all of the
    place   digits to the right of the
                            thousands place with zero.

278,362 rounded to the nearest thousand is

⟨_____.⟩

Review this example:

5. Estimate the sum by rounding each
number to the nearest hundred.

	Exact:		**Estimate:**
294	294	rounds to	300
625	625	rounds to	600
1071	1071	rounds to	1100
+ 349	+ 349	rounds to	+ 300
			2300

Estimated sum: ⟨2300⟩

Your turn:

6. Estimate the sum by rounding each
number to the nearest ten.

39
45
22
+ 17

15

Section 1.5 Rounding and Estimating

Review this example:

7. Estimate the difference by rounding each number to the nearest hundred.

	Exact:		**Estimate:**
4725	4725	rounds to	4700
− 2879	−2879	rounds to	−2900
			1800

Estimated difference: 1800

Your turn:

8. Estimate the difference by rounding each number to the nearest hundred.

1774
− 1492

Review this example:

9. In three recent months, the numbers of tons of mail that went through Hartsfield-Jackson Atlanta International Airport were 635, 687, and 567 . Round each number to the nearest hundred to estimate the tons of mail that passed through this airport.

Exact Tons of Mail:		**Estimate:**
635	rounds to	600
687	rounds to	700
+567	rounds to	+600
		1900

Estimate: approximately 1900 tons

Your turn:

10. The peak of Mt. McKinley, in Alaska, is 20,320 feet above sea level. The top of Mt. Rainier, in Washington, is 14,410 feet above sea level. Round each height to the nearest thousand to estimate the difference in elevation of these two peaks.

	Answer	Text Ref	Video Ref		Answer	Text Ref	Video Ref
1	570	Ex 1, p. 40		**6**	130		Sec 1.5, 3/5
2	640		Sec 1.5, 1/5	**7**	1800	Ex 5, p. 41	
3	278,000	Ex 2, p. 40		**8**	300		Sec 1.5, 4/5
4	36,000		Sec 1.5, 2/5	**9**	1900 tons	Ex 7, p. 42	
5	2300	Ex 4, p. 41		**10**	6000 tons		Sec 1.5, 5/5

☐ **Next, insert your homework.** Make sure you attempt all exercises asked of you and show all work, as in the exercises above. Check your answers if possible. Clearly mark any exercises you were unable to correctly complete so that you may ask questions later. DO NOT ERASE YOUR INCORRECT WORK. THIS IS HOW WE UNDERSTAND AND EXPLAIN TO YOU YOUR ERRORS.

Section 1.6 Multiplying Whole Numbers and Area

Before Class:

☐   Read the objectives on page 47.

☐   Read the **Helpful Hint** box on page 52.

☐   Read about the **Commutative Property**, **Associative Property**, and **Distributive Property** on pages 48 and 49.

☐   Complete the exercises:

1.   Use the distributive property to rewrite the expression $4(6+2)$.

2.   What is the formula for the area of a rectangle found on page 52?

3.   Is there a commutative property of subtraction? In other words, does order matter when subtracting? Why or why not?

During Class:

☐   **Write your class notes.** Neatly write down **all** examples shown as well as key terms or phrases with definitions. If not applicable or if you were absent, watch the Lecture Series (DVD) for this section and do the same (write down the examples shown as well as key terms or phrases). Insert more paper as needed.

Class Notes/Examples	Your Notes

Answers: **1)** $4 \cdot 6 + 4 \cdot 2$ **2)** area = length $\cdot$ width   **3)** There is not, because $5 - 3 \neq 3 - 5$.

Section 1.6 Multiplying Whole Numbers and Area

Class Notes (continued)

Your Notes

(Insert additional paper as needed.)

Section 1.6 Multiplying Whole Numbers and Area

Practice:

☐   Complete the Vocabulary and Readiness Check on page 55.

☐   Next, complete any incomplete exercises below. Check and correct your work using the
         answers and references at the end of this section.

Review this example:
1. Multiply.
 	
   a. 6×1
 	
   b. $0(18)$
 	
a. The product of 1 and any number is that same
    number, so $6 \times 1 = $ ⑥.
 	
b. The product of 0 and any number is 0, so
    $0(18) = $ ⓪.
	Your turn:
	2. Multiply.
	
	   a. $1 \cdot 24$
	
	   b. $0 \cdot 19$

Review this example:
3. Use the distributive property to rewrite
    the expression: $3(4 + 5)$
 	
Multiply each number inside the parentheses by 3.
 	
$3(4 + 5) = $ ⟨$3 \cdot 4 + 3 \cdot 5$⟩
	Your turn:
	4. Use the distributive property to
	    rewrite the expression: $20(14 + 6)$

Review this example: **5.** Multiply: 246×5	**Your turn:** **6.** Multiply.

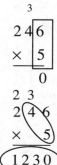

Step 1: Multiply $6 \times 5 = 30$.
Write the 0 in the ones place and
carry 3 to the tens place.

Step 2: Multiply
$4 \times 5 = 20 \, \text{tens} + 3 \, \text{tens}$
Write the 3 in the tens place and
carry 2 to the hundreds place.

Step 3: Multiply
$2 \times 5 = 10 \, \text{hundreds} + 2 \, \text{hundreds}$
Write the 2 in the hundreds place.
Carry the 1 to the thousands place.

$$\begin{array}{r} 2\,7\,7 \\ \times \quad 6 \\ \hline \end{array}$$

19

Section 1.6 Multiplying Whole Numbers and Area

Review this example:

7. The state of Colorado is in the shape of a rectangle whose length is 380 miles and whose width is 280 miles. Find its area.

$A = \text{length} \times \text{width}$
$= (380 \text{ miles})(280 \text{ miles})$
$= 106,400 \text{ square miles}$

The area of Colorado is $106,400$ square miles .

Your turn:

8. Find the area of the rectangle.

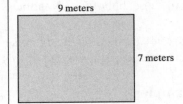

9 meters

7 meters

Complete this example:

9. A digital video disc (DVD) can hold about 4800 megabytes (MB) of information. How many megabytes can 12 DVDs hold?

megabytes per disk → 4800
×   DVDs → × 12
_____

megabytes

Your turn:

10. One tablespoon of olive oil contains 125 calories. How many calories are in 3 tablespoons of olive oil?

	Answer	Text Ref	Video Ref		Answer	Text Ref	Video Ref
1	a. 6 b. 0	Ex 1a, b, p. 48		6	1662		Sec 1.6, 4/9
2	a. 24 b. 0		Sec 1.6, 1-2/9	7	106,400 square miles	Ex 10, p. 53	
3	$3 \cdot 4 + 3 \cdot 5$	Ex 2a, p. 49		8	63 sq m		Sec 1.6, 8/9
4	$20 \cdot 14 + 20 \cdot 6$		Sec 1.6, 3/9	9	57,600 megabytes	Ex 11, p. 53	
5	1230	Ex 3b, p. 50		10	375 cal		Sec 1.6, 9/9

☐ **Next, insert your homework.** Make sure you attempt all exercises asked of you and show all work, as in the exercises above. Check your answers if possible. Clearly mark any exercises you were unable to correctly complete so that you may ask questions later. DO NOT ERASE YOUR INCORRECT WORK. THIS IS HOW WE UNDERSTAND AND EXPLAIN TO YOU YOUR ERRORS.

Section 1.7 Dividing Whole Numbers

Before Class:

☐   Read the objectives on page 60.

☐   Read the **Helpful Hint** box on page 63.

☐   Read the **Division Properties of One** box on page 61 and the **Division Properties of Zero** box on page 62.

☐   Complete the exercises:

1.   Any number divided by one is _____ .

2.   In the division problem $9\overline{)18}^{\,2}$, 18 is called the dividend, 9 is called the divisor, and 2 is called the _____ .

3.   Explain how you can use multiplication to check the division problem: $9\overline{)18}^{\,2}$

During Class:

☐   **Write your class notes.** Neatly write down **all** examples shown as well as key terms or phrases with definitions. If not applicable or if you were absent, watch the Lecture Series (DVD) for this section and do the same (write down the examples shown as well as key terms or phrases). Insert more paper as needed.

Class Notes/Examples	**Your Notes**

Answers: **1)**   that same number   **2)**   quotient   **3)** $2 \cdot 9 = 18$

Section 1.7 Dividing Whole Numbers

Class Notes (continued)	Your Notes

(Insert additional paper as needed.)

Section 1.7 Dividing Whole Numbers

Practice:

☐   Complete the Vocabulary and Readiness Check on page 69.

☐   Next, complete any incomplete exercises below. Check and correct your work using the answers and references at the end of this section.

Complete this example:
1. Fill in the blanks.

    a. $12 \div 1 = $ _____ because $12 \cdot 1 = 12$.

    b. $\dfrac{9}{9} = $ _____ because $1 \cdot 9 = 9$.

    c. $0 \div 12 = $ _____ because $0 \cdot 12 = 0$.

    d. $3 \div 0$ is _____ because a number $\cdot \ 0 \neq 3$.

Your turn:
2. Find each quotient.

    a. $31 \div 1$

    b. $\dfrac{18}{18}$

    c. $26 \div 0$ is _____ .

    d. $0 \div 14$

Review this example:
3. Divide and check: $2557 \div 7$

$$
\begin{array}{r}
365 \text{ R } 2 \\
7\overline{)2557} \\
-21 \\
\hline
45 \\
-42 \\
\hline
37 \\
-35 \\
\hline
2
\end{array}
$$

$3(7) = 21 \rightarrow 25 - 21 = 4$
Bring down the 5.

$6(7) = 42 \rightarrow 45 - 42 = 3$
Bring down the 7.

$5(7) = 35 \rightarrow 37 - 35 = 2$
The remainder is 2.

Check:

$365 \quad \cdot \quad 7 \quad + \quad 2 \quad = \quad 2557$
$\ \ \uparrow \qquad\quad \uparrow \qquad\quad \uparrow \qquad\qquad \uparrow$
whole   divisor   remainder   dividend
number   part
part

Your turn:
4. Divide and then check by multiplying.

$$55\overline{)715}$$

23

Section 1.7 Dividing Whole Numbers

Review this example:

5. Divide and check: $6819 \div 17$

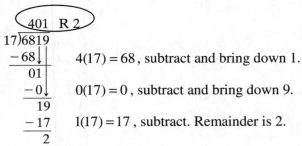

$17)\overline{6819}$

$-68\downarrow\downarrow$ $4(17) = 68$, subtract and bring down 1.

$\overline{01}$

$-0\downarrow$ $0(17) = 0$, subtract and bring down 9.

$\overline{19}$

-17 $1(17) = 17$, subtract. Remainder is 2.

$\overline{2}$

Check: $401 \cdot 17 + 2 = 6819$

Your turn:

6. Divide and check: $20,619 \div 102$

Complete this example:

7. Find the average of the numbers:

   75, 96, 81, 88

$\begin{array}{r}75\\96\\81\\+88\\\hline340\end{array}$ $4)\overline{340}$

The average is ⬭_____

Your turn:

8. Find the average of the numbers:

   86, 79, 81, 69, 80

	Answer	Text Ref	Video Ref		Answer	Text Ref	Video Ref
1	a. 12 b. 1 c. 0 d. undefined	Ex 2b, d, 3b, d, p. 61		**5**	401 R 2	Ex 8, p. 65	
2	a. 31 b. 1 c. undefined d. 0		Sec 1.7, 2&4,6&8/12	**6**	202 R15		Sec 1.7, 10/12
3	365 R 2	Ex 6, p. 64		**7**	85	Ex 12, p. 68	
4	13		Sec 1.7, 9/12	**8**	79		Sec 1.7, 12/12

☐   **Next, insert your homework.** Make sure you attempt all exercises asked of you and show all work, as in the exercises above. Check your answers if possible. Clearly mark any exercises you were unable to correctly complete so that you may ask questions later. DO NOT ERASE YOUR INCORRECT WORK. THIS IS HOW WE UNDERSTAND AND EXPLAIN TO YOU YOUR ERRORS.

Section 1.8 An Introduction to Problem Solving

Before Class:

☐   Read the objectives on page 75.

☐   Read the **Problem Solving Steps** on page 75.

☐   Read the **Key Words or Phrases of Addition, Subtraction, Multiplication, Division and Equality** chart on page 75.

☐   Complete the exercises: Fill in the chart.

	Examples	Answer
1.	What is the product of 6 and 4?	
2.	What is the total of 6 and 4?	
3.	What is 8 less than 15?	
4.	What is the quotient of 20 and 4?	
5.	What is 7 decreased by 4 less 3?	

During Class:

☐   **Write your class notes.** Neatly write down **all** examples shown as well as key terms or phrases with definitions. If not applicable or if you were absent, watch the Lecture Series (DVD) for this section and do the same (write down the examples shown as well as key terms or phrases). Insert more paper as needed.

Class Notes/Examples	**Your Notes**

Answers: **1)** $6 \cdot 4 = 24$ **2)** $6 + 4 = 10$ **3)** $15 - 8 = 7$ **4)** $20 \div 4 = 5$ **5)** $7 - 4 - 3 = 0$

Section 1.8 An Introduction to Problem Solving

Class Notes (continued)	**Your Notes**

(Insert additional paper as needed.)

Section 1.8 An Introduction to Problem Solving

Practice:

☐   Next, complete any incomplete exercises below. Check and correct your work using the answers and references at the end of this section.

Review this example:

1. The director of a learning lab at a local community college is working on next year's budget. Thirty-three new DVD players are needed at a cost of $187 each. What is the total cost of these DVD players?

UNDERSTAND:

Read and reread the problem. The cost of one DVD player is $187. To find the cost of 33 DVD players we will need to multiply. This is much quicker than doing repeated addition.

TRANSLATE:

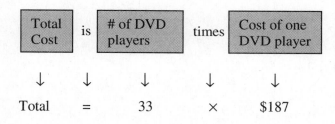

Total Cost	is	# of DVD players	times	Cost of one DVD player
↓	↓	↓	↓	↓
Total	=	33	×	$187

SOLVE:

$$\begin{array}{r} 187 \\ \times 33 \\ \hline 561 \\ 5610 \\ \hline 6171 \end{array}$$

INTERPRET:

The total cost of the DVD players is $6,171.

Your turn:

2. Find the total cost of 3 sweaters at $38 each and 5 shirts at $25 each.

UNDERSTAND:

TRANSLATE:

SOLVE:

INTERPRET:

27

Section 1.8 An Introduction to Problem Solving

Review this example:

3. The Hudson River in New York State is 306 miles long. The Snake River in the northwestern United States is 732 miles longer than the Hudson River. How long is the Snake River? (Source: U.S. Department of the Interior)

UNDERSTAND:

Read and reread the problem. The phrase "longer than" indicates we will use addition.

TRANSLATE:

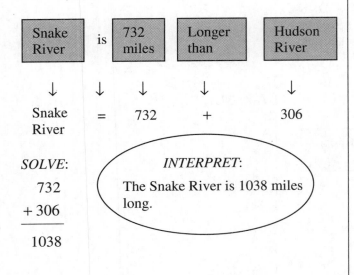

Snake River	is	732 miles	Longer than	Hudson River
↓	↓	↓	↓	↓
Snake River	=	732	+	306

SOLVE:

$$\begin{array}{r} 732 \\ + 306 \\ \hline 1038 \end{array}$$

INTERPRET:

The Snake River is 1038 miles long.

Your turn:

4. The Verrazano Narrows Bridge is the longest bridge in New York, measuring 4260 feet. The George Washington Bridge, also in New York, is 760 feet shorter than the Verrazano Narrows Bridge. Find the length of the George Washington bridge.

	Answer	Text Ref	Video Ref		Answer	Text Ref	Video Ref
1	$6171	Ex 3, p. 77		3	1038 *miles*	Ex 1, p. 75	
2	$239		Sec 1.8, 4/4	4	3500 *ft*		Sec 1.8, 3/4

☐ **Next, insert your homework.** Make sure you attempt all exercises asked of you and show all work, as in the exercises above. Check your answers if possible. Clearly mark any exercises you were unable to correctly complete so that you may ask questions later. DO NOT ERASE YOUR INCORRECT WORK. THIS IS HOW WE UNDERSTAND AND EXPLAIN TO YOU YOUR ERRORS.

Section 1.9 Exponents, Square Roots, and Order of Operations

Before Class:

☐   Read the objectives on page 85.

☐   Read the **Helpful Hint** boxes on pages 86 and 89 and the **Order of Operations** box on page 87.

☐   Complete the exercises:

    1. In $2^4 = 16$, the 2 is called the _____ and the 4 is called the _____ .

    2. To simplify $18 \div 3 \cdot 2$, which operation should be performed first?

    3. To simplify $18 \div (3 \cdot 2)$, which operation should be performed first?

During Class:

☐   **Write your class notes.** Neatly write down **all** examples shown as well as key terms or phrases with definitions. If not applicable or if you were absent, watch the Lecture Series (DVD) for this section and do the same (write down the examples shown as well as key terms or phrases). Insert more paper as needed.

Class Notes/Examples	Your Notes

Answers: **1)**   2 is the base and 4 is the exponent   **2)** $18 \div 3$ **3)** $3 \cdot 2$

Section 1.9 Exponents, Square Roots, and Order of Operations

Class Notes (continued)	**Your Notes**

(Insert additional paper as needed.)

Section 1.9 Exponents, Square Roots, and Order of Operations

Practice:

☐   Complete the Vocabulary and Readiness Check on page 91.

☐   Next, complete any incomplete exercises below. Check and correct your work using the answers and references at the end of this section.

Review this example:	**Your turn:**
1. Write using exponential notation.	**2.** Write using exponential notation.

a. $7 \cdot 7 \cdot 7$ b. $3 \cdot 3 \cdot 3 \cdot 3 \cdot 17 \cdot 17 \cdot 17$ a. $12 \cdot 12 \cdot 12$

a.   7 is a factor 3 times. The base is 7 and the
exponent is 3.

$7 \cdot 7 \cdot 7 = \boxed{7^3}$

b. $6 \cdot 6 \cdot 5 \cdot 5 \cdot 5$

b.   3 is a factor 4 times. Base = 3, Exponent = 4
17 is a factor 3 times. Base = 17, Exponent = 3

$3 \cdot 3 \cdot 3 \cdot 3 \cdot 17 \cdot 17 \cdot 17 = \boxed{3^4 \cdot 17^3}$

Complete this example:	**Your turn:**
3. Evaluate: a. 3^4 b. $5 \cdot 6^2$ c. $\sqrt{81}$	**4.** Evaluate.

a. 3^4 Write the factor 3 four times.   a. 5^3 b. 7^1
$3 \cdot 3 \cdot 3 \cdot 3 = \boxed{81}$ Then multiply.

b. $5 \cdot 6^2$
$5 \cdot 6 \cdot 6 = \bigcirc$ c. 10^2 d. $\sqrt{64}$

c. $\sqrt{81} = \bigcirc$ because ____ · ____ = 81

Complete this example:	**Your turn:**
5. Simplify: $64 \div \sqrt{64} \cdot 2 + 4$ Explain each step.	**6.** Simplify: $14 \div 7 \cdot 2 + 3$

Step 1: $64 \div 8 \cdot 2 + 4$ Step 1: _____   Explain each step.

Step 2: $8 \cdot 2 + 4$ Step 2: _____   *Step* 1:   Step 1: _____

Step 3: $16 + 4$ Step 3: _____   *Step* 2:   Step 2: _____

Step 4: $\bigcirc$ Step 4: _____   *Step* 3:   Step 3: _____

31

Section 1.9 Exponents, Square Roots, and Order of Operations

Complete this example:

7. Simplify: $\dfrac{7-2\cdot3+3^2}{5(2-1)}$

| **Your turn:** |
| **8.** Simplify: $\dfrac{7(9-6)+3}{3^2-3}$ |

The fraction bar is like a grouping symbol. We simplify above and below the fraction bar separately.

$\dfrac{7-2\cdot3+3^2}{5(2-1)} = \dfrac{7-2\cdot3+9}{5(1)}$ Evaluate 3^2 and $(2-1)$.

$= \dfrac{7-6+9}{5}$ Multiply $2\cdot3$ in the numerator. Multiply $5(1)$ in the denominator.

$= \dfrac{10}{5}$ Add/Subtract left to right.

$= \underline{\quad}$ Divide.

	Answer	Text Ref	Video Ref		Answer	Text Ref	Video Ref
1	a. 7^3 b. $3^4 \cdot 17^3$	Ex 1, 4, p. 85		**5**	20 1: Sq root 2: Divide 3: Multiply 4: Add	Ex 17, p. 89	
2	a. 12^3 b. $6^2 \cdot 5^3$		Sec 1.9, 1-2/12	**6**	7 1: Divide 2: Multiply 3: Add		Sec 1.9, 9/12
3	a. 81 b. 180 c. 9	Ex 7, 8, p. 85 Ex 10, p. 86		**7**	2	Ex 16, p. 88	
4	a. 125 b. 7 c. 100 d. 8		Sec 1.9, 3-5,7/12	**8**	4		Sec 1.9, 10/12

☐   **Next, insert your homework.** Make sure you attempt all exercises asked of you and show all work, as in the exercises above. Check your answers if possible. Clearly mark any exercises you were unable to correctly complete so that you may ask questions later. DO NOT ERASE YOUR INCORRECT WORK. THIS IS HOW WE UNDERSTAND AND EXPLAIN TO YOU YOUR ERRORS.

Preparation For Chapter 1 Test

☐   Work the Chapter 1 Vocabulary Check on page 95.

☐   Read both columns (Definitions and Concepts, and Examples) of the Chapter 1 Highlights starting on page 96.

☐   Read your Class Notes/Examples for each section covered on your Chapter 1 Test. Look for any unresolved questions you may have.

☐   Complete as many of the Chapter 1 Review exercises as possible (pages 99 - 105). Remember, the answers are in the back of your text.

☐   **Most important:** Place yourself in "test" conditions (see below) and work the Chapter 1 Test (pages 106 - 107) as a practice test the day before your actual test. To honestly assess how you are doing, try the following:

- Work on a few blank sheets of paper.
- Give yourself the same amount of time you will be given for your actual test.
- Complete this Chapter 1 Practice Test without using your notes or your text.
- If you have any time left after completing this practice test, check your work and try to find any errors on your own.
- Once done, use the back of your book to check ALL answers.
- Try to correct any errors on your own.
- Use the Chapter Test Prep Video (CTPV) to correct any errors you were unable to correct on your own. You can find these videos in the Interactive DVD Lecture Series, in MyMathLab, and on YouTube. Search MartinGayDevMath and click "Channels."

I wish you the best of luck….Elayn Martin-Gay

Section 2.1 Introduction to Fractions and Mixed Numbers

Before Class:

☐   Read the objectives on page 109.

☐   Read the **Helpful Hint** boxes on pages 109, 112, and 114.

☐   Complete the exercises:

1. Shade a part of the figure below to represent the fraction $\frac{1}{2}$.

2. Shade a part of the figure below to represent the fraction $\frac{1}{4}$.

During Class:

☐   **Write your class notes.** Neatly write down **all** examples shown as well as key terms or phrases with definitions. If not applicable or if you were absent, watch the Lecture Series (DVD) for this section and do the same (write down the examples shown as well as key terms or phrases). Insert more paper as needed.

Class Notes/Examples	**Your Notes**

Answers: 1)   2)

Section 2.1 Introduction To Fractions and Mixed Numbers

Class Notes (continued)	Your Notes

(Insert additional paper as needed.)

Section 2.1 Introduction to Fractions and Mixed Numbers

Practice:

☐   Complete the Vocabulary and Readiness Check on page 115.

☐   Next, complete any incomplete exercises below. Check and correct your work using the answers and references at the end of this section.

Review this example:	**Your turn:**
1. Identify the numerator and the denominator of each fraction.	**2.** Identify the numerator and the denominator of each fraction and identify each fraction as proper or improper.

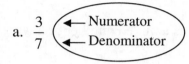

a. $\dfrac{3}{7}$ ← Numerator ← Denominator

a. $\dfrac{1}{2}$

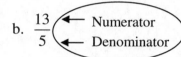

b. $\dfrac{13}{5}$ ← Numerator ← Denominator

b. $\dfrac{10}{3}$

Review this example:	**Your turn:**
3. Simplify.	**4.** Simplify.

a. $\dfrac{5}{5} = \boxed{1}$ b. $\dfrac{0}{7} = \boxed{0}$

a. $\dfrac{21}{21}$ b. $\dfrac{5}{0}$

c. $\dfrac{10}{1} = \boxed{10}$ d. $\dfrac{3}{0}$ is $\boxed{\text{undefined}}$

c. $\dfrac{13}{1}$ d. $\dfrac{0}{20}$

Complete this example:	**Your turn:**
5. Represent the shaded part of the figure as both an improper fraction and a mixed number.	**6.** Represent the shaded part of the figure as both an improper fraction and a mixed number.

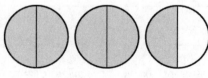

Improper Fraction: $\dfrac{5}{2}$

Mixed Number: ⬭

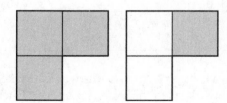

Section 2.1 Introduction To Fractions and Mixed Numbers

Review this example:	**Your turn:**
7. Write the mixed number $4\frac{2}{9}$ as an improper fraction. $4\frac{2}{9} = \frac{9\cdot4+2}{9} = \frac{36+2}{9} = \boxed{\frac{38}{9}}$	**8.** Write the mixed number $3\frac{3}{5}$ as an improper fraction.

Review this example:	**Your turn:**
9. Write $\frac{30}{7}$ as a mixed number or a whole number. Use long division: $\frac{30}{7}:\quad 7\overline{)30}^{\,4}$ $\qquad\quad\underline{28}$ $\qquad\qquad 2$ $\frac{30}{7} = \boxed{4\frac{2}{7}}$	**10.** Write $\frac{17}{5}$ as a mixed number or a whole number.

	Answer	Text Ref	Video Ref		Answer	Text Ref	Video Ref
1	a. n: 3; d: 7 b. n: 13; d: 5	Ex 1–2, p. 109		**6**	$\frac{4}{3}$; $1\frac{1}{3}$		Sec 2.1, 11/16
2	a. n: 1; d: 2; proper; b. n: 10; d: 3; improper		Sec 2.1, 1-2/16 9-10/16	**7**	$\frac{38}{9}$	Ex 17a, p. 113	
3	a. 1 b. 0 c. 10 d. undefined	Ex 3–6, p. 110		**8**	$\frac{18}{5}$		Sec 2.1, 13/16
4	a. 1 b. undefined c. 13 d. 0		Sec 2.1 3-6/16	**9**	$4\frac{2}{7}$	Ex 18a, p. 114	
5	$\frac{5}{2}$; $2\frac{1}{2}$	Ex 16, p. 113		**10**	$3\frac{2}{5}$		Sec 2.1, 15/16

☐   **Next, insert your homework.** Make sure you attempt all exercises asked of you and show all work, as in the exercises above. Check your answers if possible. Clearly mark any exercises you were unable to correctly complete so that you may ask questions later. DO NOT ERASE YOUR INCORRECT WORK. THIS IS HOW WE UNDERSTAND AND EXPLAIN TO YOU YOUR ERRORS.

Section 2.2 Factors and Prime Factorization

Before Class:

☐   Read the objectives on page 122.

☐   Read the **Helpful Hints** on pages 122, 123, 124, and 126.

☐   Use the divisibility tests to complete the exercises: (p. 124)

1. a. Is 261 divisible by 2? Yes or No   2. a. Is 960 divisible by 2? Yes or No
    b. Is 261 divisible by 3? Yes or No   b. Is 960 divisible by 3? Yes or No
    c. Is 261 divisible by 5? Yes or No   c. Is 960 divisible by 5? Yes or No

3. Circle each number below whose only factors are one and itself.

    2   3   4   5   6   7   8   9   10   11   12   13

During Class:

☐   **Write your class notes.** Neatly write down **all** examples shown as well as key terms or phrases with definitions. If not applicable or if you were absent, watch the Lecture Series (DVD) for this section and do the same (write down the examples shown as well as key terms or phrases). Insert more paper as needed.

Class Notes/Examples	Your Notes

Answers: **1)** a. no   b. yes   c. no   **2)** a. yes   b. yes   c. yes   **3)** 2, 3, 5, 7, 11, and 13

Section 2.2 Factors and Prime Factorization

Class Notes (continued)	**Your Notes**

(Insert additional paper as needed.)

Section 2.2 Factors and Prime Factorization

Practice:

☐   Complete the Vocabulary and Readiness Check on page 127.

☐   Next, complete any incomplete exercises below. Check and correct your work using the answers and references at the end of this section.

Review this example:
1. List all the factors of 20.

Write all the two-number factorizations of 20.

$1 \cdot 20 = 20$
$2 \cdot 10 = 20$
$4 \cdot 5 = 20$

The factors of 20 are 1, 2, 4, 5, 10 and 20.

Your turn:
2. List all the factors of 12.

Complete this example:
3. Determine whether each number is prime or composite. Explain your answers.

   a. 3   b. 9   c. 11   d. 17   e. 26

   a. 3 is _____. Its only factors are 1 and 3.
   b. 9 is composite. Its factors are 1, 3 and 9.
   c. 11 is prime. Its only factors are ___ and ___.
   d. 17 is _____. Its only factors are _____ and _____.
   e. 26 is composite. Its factors are 1, _____, _____, and _____.

Your turn:
4. Identify each number as prime or composite.

   a.   10   b.   67

Review this example:
5. Find the prime factorization of 180.

$3\overline{)15}$ gives 5

$3\overline{)45}$

$2\overline{)90}$

$2\overline{)180}$

Divide 180 by 2. Continue to divide by 2 until the quotient is no longer divisible by 2.

Then divide by 3 the next largest prime.

Continue to divide by 3 until the quotient is no longer divisible by 3.

Continue process until the quotient is a prime number.

Prime factorization: $2 \cdot 2 \cdot 3 \cdot 3 \cdot 5 = 2^2 \cdot 3^2 \cdot 5$

Your turn:
6. Find the prime factorization of 36. Write repeated factors using exponents.

41

Section 2.2 Factors and Prime Factorization

Review this example:

7. Use a factor tree to find the prime factorization of 80.

Write 80 as a product of two numbers. Continue factoring each number until all factors are prime.

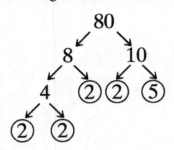

Prime factorization: $2 \cdot 2 \cdot 2 \cdot 2 \cdot 5 = 2^4 \cdot 5$

Your turn:

8. Write the prime factorization of 240.

	Answer	Text Ref	Video Ref		Answer	Text Ref	Video Ref
1	1, 2, 4, 5, 10, and 20	Ex 1, p. 122		5	$2 \cdot 2 \cdot 3 \cdot 3 \cdot 5 =$ $2^2 \cdot 3^2 \cdot 5$	Ex 4, p. 124	
2	1, 2, 3, 4, 6, and 12		Sec 2.2, 2/8	6	$2^2 \cdot 3^2$		Sec 2.2, 7/8
3	a. prime c. 1, 11 d. 17 is prime; 1, 17 e. 2, 13, and 26	Ex 2, p. 123		7	$2 \cdot 2 \cdot 2 \cdot 2 \cdot 5 = 2^4 \cdot 5$	Ex 7, p. 126	
4	a. composite b. prime		Sec 2.2, 4-5/8	8	$2 \cdot 2 \cdot 2 \cdot 2 \cdot 3 \cdot 5 =$ $2^4 \cdot 3 \cdot 5$		Sec 2.2, 8/8

☐   **Next, insert your homework.** Make sure you attempt all exercises asked of you and show all work, as in the exercises above. Check your answers if possible. Clearly mark any exercises you were unable to correctly complete so that you may ask questions later. DO NOT ERASE YOUR INCORRECT WORK. THIS IS HOW WE UNDERSTAND AND EXPLAIN TO YOU YOUR ERRORS.

Section 2.3 Simplest Form of a Fraction

Before Class:

☐ Read the objectives on page 129.

☐ Read the **Helpful Hint** boxes on pages 130, 131, 133, and 134.

☐ Complete the exercises:

1. Simplify the following fractions.

   a. $\dfrac{28}{28}$ b. $\dfrac{28}{0}$ c. $\dfrac{0}{28}$

2. Write the prime factorization of 80 using division.

3. Use a factor tree to find the prime factorization of 700.

During Class:

☐ **Write your class notes.** Neatly write down **all** examples shown as well as key terms or phrases with definitions. If not applicable or if you were absent, watch the Lecture Series (DVD) for this section and do the same (write down the examples shown as well as key terms or phrases). Insert more paper as needed.

Class Notes/Examples	Your Notes

Answers: **1)** a. 1   b. undefined   c. 0   **2)** $2^4 \cdot 5$ **3)** $2^2 \cdot 5^2 \cdot 7$

Section 2.3 Simplest Form of a Fraction

Class Notes (continued)	**Your Notes**

(Insert additional paper as needed.)

Section 2.3 Simplest Form of a Fraction

Practice:

☐   Complete the Vocabulary and Readiness Check on page 135.

☐   Next, complete any incomplete exercises below. Check and correct your work using the answers and references at the end of this section.

Review this example:	**Your turn:**
1. Write the fraction $\dfrac{12}{20}$ in simplest form.	**2.** Write the fraction $\dfrac{14}{16}$ in simplest form.
Notice that 12 and 20 have a common factor of 4.	
$\dfrac{12}{20}=\dfrac{4\cdot 3}{4\cdot 5}=\dfrac{4}{4}\cdot\dfrac{3}{5}=1\cdot\dfrac{3}{5}=\boxed{\dfrac{3}{5}}$ 3 and 5 have no common factors (other than 1)	
Review this example:	**Your turn:**
3. Write the fraction $\dfrac{42}{66}$ in simplest form. You may also write the prime factorization of the numerator and the denominator.	**4.** Write the fraction $\dfrac{24}{40}$ in simplest form using prime factorization.
$\dfrac{42}{66}=\dfrac{2\cdot 3\cdot 7}{2\cdot 3\cdot 11}=\dfrac{2}{2}\cdot\dfrac{3}{3}\cdot\dfrac{7}{11}=1\cdot 1\cdot\dfrac{7}{11}=\boxed{\dfrac{7}{11}}$	
Review this example:	**Your turn:**
5. Write the fraction $\dfrac{6}{60}$ in simplest form.	**6.** Write the fraction $\dfrac{70}{196}$ in simplest form.
We can use a shortcut procedure with common factors when simplifying. Dividing out a common factor in the numerator and denominator is the same as removing a factor of 1 in the product.	
$\dfrac{6}{60}=\dfrac{\overset{1}{\cancel{2}}\cdot\overset{1}{\cancel{3}}}{\underset{1}{\cancel{2}}\cdot 2\cdot\underset{1}{\cancel{3}}\cdot 5}=\dfrac{1\cdot 1}{1\cdot 2\cdot 1\cdot 5}=\boxed{\dfrac{1}{10}}$	

45

Section 2.3 Simplest Form of a Fraction

Review this example:

7. Determine whether $\dfrac{16}{40}$ and $\dfrac{10}{25}$ are equivalent.

$$\frac{16}{40} = \frac{\overset{1}{\cancel{8}} \cdot 2}{\underset{1}{\cancel{8}} \cdot 5} = \frac{1 \cdot 2}{1 \cdot 5} = \boxed{\frac{2}{5}}$$ 16 and 40 have a common factor of 8.

$$\frac{10}{25} = \frac{\overset{1}{\cancel{5}} \cdot 2}{\underset{1}{\cancel{5}} \cdot 5} = \frac{1 \cdot 2}{1 \cdot 5} = \boxed{\frac{2}{5}}$$ 10 and 25 have a common factor of 5.

The two fractions simplify to the same number.

Thus, $\boxed{\dfrac{16}{40} = \dfrac{10}{25}}$.

Your turn:

8. Determine whether $\dfrac{3}{9}$ and $\dfrac{6}{18}$ are equivalent.

	Answer	Text Ref	Video Ref		Answer	Text Ref	Video Ref
1	$\dfrac{3}{5}$	Ex 1, p. 130		**5**	$\dfrac{1}{10}$	Ex 6, p. 131	
2	$\dfrac{7}{8}$		Sec 2.3, 1/7	**6**	$\dfrac{5}{14}$		Sec 2.3, 3/7
3	$\dfrac{7}{11}$	Ex 2, p. 130		**7**	equivalent	Ex 8, p. 132	
4	$\dfrac{3}{5}$		Sec 2.3, 2/7	**8**	equivalent		Sec 2.3, 5/7

☐ **Next, insert your homework.** Make sure you attempt all exercises asked of you and show all work, as in the exercises above. Check your answers if possible. Clearly mark any exercises you were unable to correctly complete so that you may ask questions later. DO NOT ERASE YOUR INCORRECT WORK. THIS IS HOW WE UNDERSTAND AND EXPLAIN TO YOU YOUR ERRORS.

Section 2.4 Multiplying Fractions and Mixed Numbers

Before Class:

☐ Read the objectives on page 141.

☐ Read the **Helpful Hint** boxes on pages 142, 144, and 145.

☐ Complete the exercises:

1. Simplify each fraction.

     a. $\dfrac{12}{18}$ b. $\dfrac{15}{30}$

2. Determine whether each fraction is equivalent to $\dfrac{2}{5}$.

     a. $\dfrac{8}{20}$ b. $\dfrac{10}{25}$ c. $\dfrac{6}{10}$

During Class:

☐ **Write your class notes.** Neatly write down **all** examples shown as well as key terms or phrases with definitions. If not applicable or if you were absent, watch the Lecture Series (DVD) for this section and do the same (write down the examples shown as well as key terms or phrases). Insert more paper as needed.

Class Notes/Examples	**Your Notes**

Answers: **1)** a. $\dfrac{2}{3}$ b. $\dfrac{1}{2}$ **2)** a. yes   b. yes   c. no

Section 2.4 Multiplying Fractions and Mixed Numbers

Class Notes (continued)	**Your Notes**

(Insert additional paper as needed.)

Section 2.4 Multiplying Fractions and Mixed Numbers

Practice:

☐   Complete the Vocabulary and Readiness Check on page 146.

☐   Next, complete any incomplete exercises below. Check and correct your work using the answers and references at the end of this section.

Review this example:

1. Multiply $\dfrac{2}{3} \cdot \dfrac{5}{11}$. Write your answer in simplest form.

$$\frac{2}{3} \cdot \frac{5}{11} = \frac{2 \cdot 5}{3 \cdot 11} = \boxed{\frac{10}{33}}$$ Multiply numerators.

Multiply denominators.

This fraction is simplified since 10 and 33 have no common factors.

Your turn:

2. Multiply $\dfrac{6}{5} \cdot \dfrac{1}{7}$. Write your answer in simplest form.

Review this example:

3. Multiply $\dfrac{6}{7} \cdot \dfrac{14}{27}$. Write your answer in simplest form.

We can simplify by finding the prime factorizations and then divide out common factors in the numerator and denominator.

$$\frac{6}{7} \cdot \frac{14}{27} = \frac{2 \cdot \overset{1}{\cancel{3}} \cdot 2 \cdot \overset{1}{\cancel{7}}}{\underset{1}{\cancel{7}} \cdot \underset{1}{\cancel{3}} \cdot 3 \cdot 3} = \frac{2 \cdot 2}{3 \cdot 3} = \boxed{\frac{4}{9}}$$

Your turn:

4. Multiply $\dfrac{2}{7} \cdot \dfrac{5}{8}$. Write your answer in simplest form.

Review this example:

5. Multiply $\dfrac{3}{4} \cdot 20$. Write your answer in simplest form.

Recall $20 = \dfrac{20}{1}$.

$$\frac{3}{4} \cdot 20 = \frac{3}{4} \cdot \frac{20}{1} = \frac{3 \cdot \overset{5}{\cancel{4}} \cdot 5}{\underset{1}{\cancel{4}} \cdot 1} = \frac{15}{1} = \boxed{15}$$

Your turn:

6. Multiply $\dfrac{5}{8} \cdot 4$. Write your answer in simplest form.

Section 2.4 Multiplying Fractions and Mixed Numbers

Review this example:

7. Multiply $1\frac{2}{3} \cdot 2\frac{1}{4}$. Check by estimating.

Write any mixed or whole number as improper fractions. Then multiply.

Exact: $1\frac{2}{3} \cdot 2\frac{1}{4} = \frac{5}{3} \cdot \frac{9}{4} = \frac{5 \cdot 9}{3 \cdot 4} = \frac{5 \cdot \overset{1}{\cancel{3}} \cdot 3}{\underset{1}{\cancel{3}} \cdot 4} = \boxed{\frac{15}{4} \; or \; 3\frac{3}{4}}$

Estimate: $1\frac{2}{3}$ rounds to 2, $2\frac{1}{4}$ rounds to 2, and
$2 \cdot 2 = \boxed{4}$

The estimate is close to the exact value, so our answer is reasonable.

Your turn:

8. Multiply $2\frac{1}{5} \cdot 3\frac{1}{2}$. Check by estimating.

	Answer	Text Ref	Video Ref		Answer	Text Ref	Video Ref
1	$\frac{10}{33}$	Ex 1, p. 141		**5**	15	Ex 9, p. 143	
2	$\frac{6}{35}$		Sec 2.4, 1/7	**6**	$\frac{5}{2} \; or \; 2\frac{1}{2}$		Sec 2.4, 4/7
3	$\frac{4}{9}$	Ex 3, p. 142		**7**	$\frac{15}{4} \; or \; 3\frac{3}{4}$;	Ex 10, p. 144	
4	$\frac{5}{28}$		Sec 2.4, 2/7	**8**	$\frac{77}{10} \; or \; 7\frac{7}{10}$;		Sec 2.4, 5/7

☐ **Next, insert your homework.** Make sure you attempt all exercises asked of you and show all work, as in the exercises above. Check your answers if possible. Clearly mark any exercises you were unable to correctly complete so that you may ask questions later. DO NOT ERASE YOUR INCORRECT WORK. THIS IS HOW WE UNDERSTAND AND EXPLAIN TO YOU YOUR ERRORS.

Section 2.5 Dividing Fractions and Mixed Numbers

Before Class:

☐   Read the objectives on page 152.

☐   Read the **Helpful Hint** boxes on pages 152 and 153.

☐   Complete the exercises:

Find the error in each calculation.

1. $3\frac{2}{3}\cdot 1\frac{1}{7}=3\frac{2}{21}$

2. $5\cdot 2\frac{1}{4}=10\frac{1}{4}$

During Class:

☐   **Write your class notes.** Neatly write down **all** examples shown as well as key terms or phrases with definitions. If not applicable or if you were absent, watch the Lecture Series (DVD) for this section and do the same (write down the examples shown as well as key terms or phrases). Insert more paper as needed.

Class Notes/Examples	**Your Notes**

Answers: **1)** $3\frac{2}{3}\cdot 1\frac{1}{7}=\frac{11}{3}\cdot\frac{8}{7}=\frac{88}{21}=4\frac{4}{21}$ **2)** $5\cdot 2\frac{1}{4}=\frac{5}{1}\cdot\frac{9}{4}=\frac{45}{4}=11\frac{1}{4}$

Section 2.5 Dividing Fractions and Mixed Numbers

Class Notes (continued)	Your Notes

(Insert additional paper as needed.)

Section 2.5 Dividing Fractions and Mixed Numbers

Practice:

☐   Complete the Vocabulary and Readiness Check on page 156.

☐   Next, complete any incomplete exercises below. Check and correct your work using the answers and references at the end of this section.

Review this example:
1. Find the reciprocal of each number.

a. $\dfrac{5}{6}$ b. 5

To find the reciprocal of a fraction, interchange its numerator and denominator.

a. $\left(\dfrac{6}{5}\right)$ because $\dfrac{5}{6}\cdot\dfrac{6}{5}=\dfrac{5\cdot6}{6\cdot5}=\dfrac{30}{30}=1$

b. $\left(\dfrac{1}{5}\right)$ because $\dfrac{5}{1}\cdot\dfrac{1}{5}=\dfrac{5\cdot1}{1\cdot5}=\dfrac{5}{5}=1$

Your turn:
2. Find the reciprocal of each number.

a. $\dfrac{4}{7}$ b. 15

Complete this example:
3. Divide and simplify:

a. $\dfrac{7}{8}\div\dfrac{2}{9}$ b. $\dfrac{5}{16}\div\dfrac{3}{4}$

Multiply the first fraction by the reciprocal of the second fraction.

a. $\dfrac{7}{8}\div\dfrac{2}{9}=\dfrac{7}{8}\cdot\dfrac{9}{2}=\dfrac{7\cdot9}{8\cdot2}=\left(\dfrac{63}{16}\right)$

b. $\dfrac{5}{16}\div\dfrac{3}{4}=\dfrac{5}{16}\cdot\dfrac{4}{3}=\dfrac{5\cdot4}{16\cdot3}=\underline{\qquad}=\bigcirc$

Your turn:
4. Divide and simplify:

a. $\dfrac{2}{3}\div\dfrac{5}{6}$ b. $\dfrac{3}{25}\div\dfrac{27}{40}$

53

©2011 Pearson Education, Inc. Publishing as Prentice Hall

Section 2.5 Dividing Fractions and Mixed Numbers

Review this example:

5. Divide and simplify: $\dfrac{3}{4} \div 5$

Write any mixed or whole number as improper fractions. Then divide as usual.

$$\dfrac{3}{4} \div 5 = \dfrac{3}{4} \div \dfrac{5}{1} = \dfrac{3}{4} \cdot \dfrac{1}{5} = \dfrac{3 \cdot 1}{4 \cdot 5} = \boxed{\dfrac{3}{20}}$$

Your turn:

6. Divide and simplify: $\dfrac{2}{3} \div 4$

Review this example:

7. Divide and simplify: $5\dfrac{2}{3} \div 2\dfrac{5}{9}$

$$5\dfrac{2}{3} \div 2\dfrac{5}{9} = \dfrac{17}{3} \div \dfrac{23}{9} = \dfrac{17}{3} \cdot \dfrac{9}{23} = \dfrac{17 \cdot 9}{3 \cdot 23} = \dfrac{17 \cdot \overset{1}{\cancel{3}} \cdot 3}{\underset{1}{\cancel{3}} \cdot 23} = \boxed{\dfrac{51}{23} \; or \; 2\dfrac{5}{23}}$$

Your turn:

8. Divide and simplify:

$3\dfrac{3}{7} \div 3\dfrac{1}{3}$

Review this example:

9. Divide and simplify: $0 \div \dfrac{2}{21}$

$$0 \div \dfrac{2}{21} = 0 \cdot \dfrac{21}{2} = \boxed{0}$$

Recall that 0 multiplied by any number is 0.

Your turn:

10. Divide and simplify:

$\dfrac{8}{13} \div 0$

	Answer	Text Ref	Video Ref		Answer	Text Ref	Video Ref
1	a. $\dfrac{6}{5}$ b. $\dfrac{1}{5}$	Ex 1&4, p. 152		**6**	$\dfrac{1}{6}$		Sec 2.5, 7/9
2	a. $\dfrac{7}{4}$ b. $\dfrac{1}{15}$		Sec 2.5, 1-2/9	**7**	$\dfrac{51}{23}$ or $2\dfrac{5}{23}$	Ex 10, p. 154	
3	a. $\dfrac{63}{16}$ b. $\dfrac{5}{12}$	Ex 5&6, p. 153		**8**	$\dfrac{36}{35}$ or $1\dfrac{1}{35}$		Sec 2.5, 8/9
4	a. $\dfrac{4}{5}$ b. $\dfrac{8}{45}$		Sec 2.5, 3&5/9	**9**	0	Ex 11, p. 154	
5	$\dfrac{3}{20}$	Ex 8, p. 154		**10**	undefined		Sec 2.5, 6/9

☐   **Next, insert your homework.** Make sure you attempt all exercises asked of you and show all work, as in the exercises above. Check your answers if possible. Clearly mark any exercises you were unable to correctly complete so that you may ask questions later. DO NOT ERASE YOUR INCORRECT WORK. THIS IS HOW WE UNDERSTAND AND EXPLAIN TO YOU YOUR ERRORS.

Preparing for the Chapter 2 Test

Start preparing for your Chapter 2 Test as soon as possible. Pay careful attention to any instructor discussion about this test, especially discussion on what sections you will be responsible for, etc.

☐ Work the Chapter 2 Vocabulary Check on page 161.

☐ Read both columns (Definitions and Concepts, and Examples) of the Chapter 2 Highlights starting on page 161.

☐ Read your Class Notes/Examples for each section covered on your Chapter 2 Test. Look for any unresolved questions you may have.

☐ Complete as many of the Chapter 2 Review exercises as possible (starting on page 164). Remember, the answers are in the back of your text.

☐ **Most important:** Place yourself in "test" conditions (see below) and work the Chapter 2 Test (pages 168 – 169) as a practice test the day before your actual test. To honestly assess how you are doing, try the following:

- Work on a few blank sheets of paper.
- Give yourself the same amount of time you will be given for your actual test.
- Complete this Chapter 2 Practice Test without using your notes or your text.
- If you have any time left after completing this practice test, check your work and try to find any errors on your own.
- Once done, use the back of your book to check ALL answers.
- Try to correct any errors on your own.
- Use the Chapter Test Prep Video (CTPV) to correct any errors you were unable to correct on your own. You can find these videos in the Interactive DVD Lecture Series, in MyMathLab, and on YouTube. Search MartinGayDevMath and click "Channels."

I wish you the best of luck....Elayn Martin-Gay

Section 3.1 Adding and Subtracting Like Fractions

Before Class:

☐   Read the objectives on page 174.

☐   Read the **Helpful Hint** box on page 174.

☐   Complete the exercises:

1. Add: $\dfrac{1}{2}+\dfrac{1}{2}=$ _____   2. Add: $\dfrac{2}{3}+\dfrac{1}{3}=$ _____   3. Simplify: $\dfrac{16}{24}=$ _____

4. Find the perimeter:

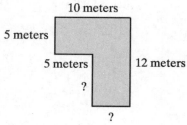

During Class:

☐   **Write your class notes.** Neatly write down **all** examples shown as well as key terms or phrases with definitions. If not applicable or if you were absent, watch the Lecture Series (DVD) for this section and do the same (write down the examples shown as well as key terms or phrases). Insert more paper as needed.

Class Notes/Examples	**Your Notes**

Answers: **1)** 1   **2)** 1   **3)** $\dfrac{2}{3}$ **4)** 44 *m*

Section 3.1 Adding and Subtracting Like Fractions

Class Notes (continued)	**Your Notes**

(Insert additional paper as needed.)

Section 3.1 Adding and Subtracting Like Fractions

Practice:

☐   Complete the Vocabulary and Readiness Check on page 178.

☐   Next, complete any incomplete exercises below. Check and correct your work using the answers and references at the end of this section.

Complete this example:	**Your turn:**

1. a. Add and simplify. $\dfrac{2}{7}+\dfrac{3}{7}$

The common denominator is 7.   Add the numerators and keep the common denominator.

$\dfrac{2}{7}+\dfrac{3}{7}=\left(\underline{}\right)$

     b. Add and simplify. $\dfrac{3}{16}+\dfrac{7}{16}$

The common denominator is 16.   Add the numerators and keep the common denominator.

$\dfrac{3}{16}+\dfrac{7}{16}=\dfrac{10}{16}=\left(\underline{}\right)$ Write the fraction in simplified form.

2. a. Add and simplify. $\dfrac{2}{9}+\dfrac{4}{9}$

     b. Add and simplify. $\dfrac{4}{13}+\dfrac{2}{13}+\dfrac{1}{13}$

Complete this example:	**Your turn:**

3. Subtract and simplify. $\dfrac{7}{8}-\dfrac{5}{8}$

The common denominator is 8. Subtract the numerators and keep the common denominator.

$\dfrac{7}{8}-\dfrac{5}{8}=\dfrac{7-5}{8}=\left(\underline{}=\underline{}\right)$

4. Subtract and simplify. $\dfrac{7}{8}-\dfrac{1}{8}$

Section 3.1 Adding and Subtracting Like Fractions

Complete this example:

5. Find the perimeter of the rectangle.

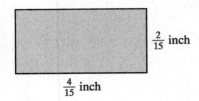

Recall that in a rectangle opposite sides are equal. Add all four sides to obtain the perimeter.

$$\text{Perimeter} = \frac{2}{15} + \frac{4}{15} + \underline{\hspace{1.5cm}} + \underline{\hspace{1.5cm}} = \frac{}{15}$$

The perimeter in simplified form = $\left(\dfrac{}{15} \; in. \right)$

Your turn:

6. Find the perimeter of the triangle.

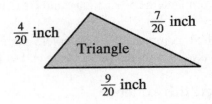

	Answer	Text Ref	Video Ref		Answer	Text Ref	Video Ref
1	a. $\frac{5}{7}$ b. $\frac{5}{8}$	Ex 1, 2, p. 175		**4**	$\frac{3}{4}$		Sec 3.1, 5/7
2	a. $\frac{2}{3}$ b. $\frac{7}{13}$		Sec 3.1, 2 - 3/7	**5**	$\frac{12}{15} = \frac{4}{5}$ inch	Ex 6, p. 176	
3	$\frac{2}{8} = \frac{1}{4}$	Ex 5, p. 175		**6**	1 inch		Sec 3.1, 6/7

☐   **Next, insert your homework.** Make sure you attempt all exercises asked of you and show all work, as in the exercises above. Check your answers if possible. Clearly mark any exercises you were unable to correctly complete so that you may ask questions later. DO NOT ERASE YOUR INCORRECT WORK. THIS IS HOW WE UNDERSTAND AND EXPLAIN TO YOU YOUR ERRORS.

Section 3.2 Least Common Multiple

Before Class:

☐   Read the objectives on page 183.

☐   Read the **Helpful Hint** boxes on pages 185 and 187.

☐   Complete the exercises:

1.   The first eight multiples of 5 are 5, 10, 15, 20, 25, 30, 35, 40, . . .
   Write down the first eight multiples of three.

2.   Multiples of 5:   5, 10, 15, 20, 25, 30, 35, 40, . . .
   Multiples of 8:   8, 16, 24, 32, 40, 48, 56, 64, . . .

   What is the least common multiple of 5 and 8? _____

3.   Write the prime factorization of 24.

During Class:

☐   **Write your class notes.** Neatly write down **all** examples shown as well as key terms or phrases with definitions. If not applicable or if you were absent, watch the Lecture Series (DVD) for this section and do the same (write down the examples shown as well as key terms or phrases.) Insert more paper as needed.

Class Notes/Examples	**Your Notes**

Answers: **1)**   3, 6, 9, 12, 15, 18, 21, 24   **2)**   40   **3)** $24 = 2^3 \cdot 3$

Section 3.2 Least Common Multiple

Class Notes (continued)	**Your Notes**

(Insert additional paper as needed.)

Practice:

☐   Complete the Vocabulary and Readiness Check on page 188.

☐   Next, complete any incomplete exercises below. Check and correct your work using the answers and references at the end of this section.

Review this example:
1. Find the LCM of 9 and 12.

Method 1: Find the LCM using multiples.

List multiples of each number until you see a common multiple.

Multiples of 12: 12 , 24 , 36 , 48 , 60 , . . .

Multiples of 9:   9 , 18 , 27 , 36 , 45 , . . .

The LCM of 9 and 12 is 36.

Method 2: Find the LCM using prime factorization.

Prime factors of $12 = 2^2 \cdot 3$

Prime factors of $9 = 3^2$

For each different prime factor, circle the greatest number of times that factor occurs in any one factorization.

$12 = \boxed{2^2} \cdot 3$

$9 = \boxed{3^2}$

The LCM is the product of the circled factors.

Thus, the LCM of 9 and 12 is

$2^2 \cdot 3^2 = 4 \cdot 9 = 36$

Your turn:
2. Find the LCM of 9 and 15.

Method 1: Find the LCM using multiples.

Method 2: Find the LCM using prime factorization.

63

Section 3.2 Least Common Multiple

Review this example:

3. Find the LCM of 15, 18 and 54. Use prime factors.

Prime factors of $15 = 3 \cdot \boxed{5}$

Prime factors of $18 = \boxed{2} \cdot 3^2$

Prime factors of $54 = 2 \cdot \boxed{3^3}$

The LCM of 15, 18 and 54 is

$2 \cdot 3^3 \cdot 5 = 2 \cdot 27 \cdot 5 = 270$

Your turn:

4. Find the LCM of 30, 36 and 50. Use prime factors.

Review this example:

5. Write an equivalent fraction with the indicated denominator.

$$\frac{3}{4} = \frac{}{20}$$

Since $4 \cdot 5 = 20$ multiply $\frac{3}{4}$ by $\frac{5}{5}$.

$$\frac{3}{4} \cdot \frac{5}{5} = \frac{3 \cdot 5}{4 \cdot 5} = \boxed{\frac{15}{20}}$$

Your turn:

6. Write an equivalent fraction with the indicated denominator.

$$\frac{4}{7} = \frac{}{35}$$

	Answer	Text Ref	Video Ref		Answer	Text Ref	Video Ref
1	36	Ex 2, p. 183		4	900		Sec 3.2, 4/6
2	45		Sec 3.2, 1/6	5	$\frac{15}{20}$	Ex 8, p. 186	
3	270	Ex 6, p. 185		6	$\frac{20}{35}$		Sec 3.2, 5/6

☐ **Next, insert your homework.** Make sure you attempt all exercises asked of you and show all work, as in the exercises above. Check your answers if possible. Clearly mark any exercises you were unable to correctly complete so that you may ask questions later. DO NOT ERASE YOUR INCORRECT WORK. THIS IS HOW WE UNDERSTAND AND EXPLAIN TO YOU YOUR ERRORS.

Section 3.3 Adding and Subtracting Unlike Fractions

Before Class:

☐   Read the objectives on page 191.

☐   Complete the exercises:

1.   Find the LCD of the following two fractions: $\dfrac{1}{18}, \dfrac{3}{4}$.

2.   Write each fraction as an equivalent fraction with the given denominator.

   a. $\dfrac{5}{14} = \dfrac{}{42}$ b. $\dfrac{7}{9} = \dfrac{}{45}$ c. $\dfrac{3x}{4} = \dfrac{}{48}$

During Class:

☐   **Write your class notes.** Neatly write down **all** examples shown as well as key terms or phrases with definitions. If not applicable or if you were absent, watch the Lecture Series (DVD) for this section and do the same (write down the examples shown as well as key terms or phrases.) Insert more paper as needed.

Class Notes/Example	**Your Notes**

Answers: **1)** 36   **2)** a. 15   b. 35   c. 36x

Section 3.3 Adding and Subtracting Unlike Fractions

Class Notes (continued)	**Your Notes**

(Insert additional paper as needed.)

Section 3.3 Adding and Subtracting Unlike Fractions

Practice:

☐   Complete the Vocabulary and Readiness Check on page 197.

☐   Next, complete any incomplete exercises below. Check and correct your work using the answers and references at the end of this section.

Review this example:

1. Add: $\dfrac{2}{5} + \dfrac{4}{15}$

Step 1: Find the LCD of the fractions.

    The LCD is 15.

Step 2: Write each fraction as an equivalent fraction whose denominator is the LCD.

$$\frac{2}{5} = \frac{2}{5} \cdot \frac{3}{3} = \frac{6}{15}, \quad \frac{4}{15} = \frac{4}{15}$$

Step 3: Add or subtract the like fractions.

$$\frac{2}{5} + \frac{4}{15} = \frac{6}{15} + \frac{4}{15} = \frac{10}{15}$$

Step 4: Write in simplest form.

$$\frac{10}{15} = \frac{2 \cdot \cancel{5}^{1}}{3 \cdot \cancel{5}_{1}} = \boxed{\frac{2}{3}}$$

Your turn:

2. Add: $\dfrac{7}{15} + \dfrac{5}{12}$

Review this example:

3. Subtract: $\dfrac{10}{11} - \dfrac{2}{3}$

Step 1: The LCD of 3 and 11 is 33.

Step 2: $\dfrac{10}{11} = \dfrac{10}{11} \cdot \dfrac{3}{3} = \dfrac{30}{33}$ and $\dfrac{2}{3} = \dfrac{2}{3} \cdot \dfrac{11}{11} = \dfrac{22}{33}$

Step 3: $\dfrac{10}{11} - \dfrac{2}{3} = \dfrac{30}{33} - \dfrac{22}{33} = \dfrac{8}{33}$ or $\dfrac{8}{33}$

Step 4: $\boxed{\dfrac{8}{33}}$ is in simplest form.

Your turn:

4. Subtract: $\dfrac{5}{6} - \dfrac{3}{7}$

Section 3.3 Adding and Subtracting Unlike Fractions

Complete this example:

5. A freight truck has $\frac{1}{4}$ ton of computers, $\frac{1}{3}$ ton of televisions, and $\frac{3}{8}$ ton of small appliances. Find the total weight of its load.

UNDERSTAND and TRANSLATE:

Reread the problem and look for key words. To find the

[total] weight means to add the weights.

$\frac{1}{4}+\frac{1}{3}+\frac{3}{8}$ The LCD is 24.

SOLVE: $\frac{1}{4}\cdot\frac{6}{6}+\frac{1}{3}\cdot\frac{8}{8}+\frac{3}{8}\cdot\frac{3}{3}=\frac{6}{24}+\frac{8}{24}+\frac{9}{24}=\left(\underline{\quad}\right)$

INTERPRET:

The total weight of the truck's load is $\left(\underline{\quad}\right)$ ton.

Your turn:

6. Find the perimeter of the geometric figure.

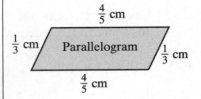

	Answer	Text Ref	Video Ref		Answer	Text Ref	Video Ref
1	$\frac{2}{3}$	Ex 1, p. 192		**4**	$\frac{17}{42}$		Sec 3.3, 3/5
2	$\frac{53}{60}$		Sec 3.3, 2/5	**5**	$\frac{23}{24}$ ton	Ex 8, p. 194	
3	$\frac{8}{33}$	Ex 6, p. 193		**6**	$\frac{34}{15}$ or $2\frac{4}{15}$ cm		Sec 3.3, 5/5

☐ **Next, insert your homework.** Make sure you attempt all exercises asked of you and show all work, as in the exercises above. Check your answers if possible. Clearly mark any exercises you were unable to correctly complete so that you may ask questions later. DO NOT ERASE YOUR INCORRECT WORK. THIS IS HOW WE UNDERSTAND AND EXPLAIN TO YOU YOUR ERRORS.

Section 3.4 Adding and Subtracting Mixed Numbers

Before Class:

☐   Read the objectives on page 204.

☐   Read the **Helpful Hint** box on page 205.

☐   Complete the exercises:

    1. Which of the following are equivalent to 8?

        a. $5\dfrac{9}{3}$ b. $7\dfrac{9}{9}$ c. $\dfrac{16}{2}$ d. all of these

    2. Write $5\dfrac{1}{3}$ as an improper fraction. Is $5\dfrac{1}{3}$ closer to 5 or closer to 6?

    3. Write $\dfrac{8}{3}$ as a mixed number. Is $\dfrac{8}{3}$ closer to 2 or closer to 3?

During Class:

☐   **Write your class notes.** Neatly write down **all** examples shown as well as key terms or phrases with definitions. If not applicable or if you were absent, watch the Lecture Series (DVD) for this section and do the same (write down the examples shown as well as key terms or phrases). Insert more paper as needed.

Class Notes/Examples	**Your Notes**

Answers: **1)** all of these   **2)** $\dfrac{16}{3}$ is closer to 5   **3)** $2\dfrac{2}{3}$ is closer to 3

Section 3.4 Adding and Subtracting Mixed Numbers

Class Notes (continued)	**Your Notes**

(Insert additional paper as needed.)

Section 3.4 Adding and Subtracting Mixed Numbers

Practice:

☐   Complete the Vocabulary and Readiness Check on page 209.

☐   Complete any incomplete exercises below. Check and correct your work using the
answers and references at the end of this section.

Review this example:

1. Add: $2\dfrac{1}{3}+5\dfrac{3}{8}$. Check by estimating.

The LCD of 3 and 8 is 24.

$$2\dfrac{1\cdot 8}{3\cdot 8}=2\dfrac{8}{24}$$
$$+5\dfrac{3\cdot 3}{8\cdot 3}=5\dfrac{9}{24}$$

Add the fractions.
Add the whole
numbers.

$$7\dfrac{17}{24}$$

Estimate: $2\dfrac{1}{3}$ is close to 2 and $5\dfrac{3}{8}$ is close to
5. Add $2+5=7$ which is close to our answer.

Your turn:

2. Add: $10\dfrac{3}{14}+3\dfrac{4}{7}$. Check by
estimating.

Complete this example:

3. Add: $3\dfrac{4}{5}+1\dfrac{4}{15}$

The LCD of 5 and 15 is 15.

$$3\dfrac{4}{5}=3\dfrac{12}{15}$$
$$+1\dfrac{4}{15}=1\dfrac{4}{15}$$

$$4\dfrac{16}{15}$$

Notice the fraction part
is improper.

Rewrite the improper fraction as a mixed number.
Then, add the whole numbers together.

$\dfrac{16}{15}$ is $1\dfrac{1}{15}$, so $4\dfrac{16}{15}=4+1\dfrac{1}{15}=$ (_____ .

Your turn:

4. Add: $1\dfrac{5}{6}+5\dfrac{3}{8}$.

Section 3.4 Adding and Subtracting Mixed Numbers

Review this example:

5. Subtract: $12 - 8\dfrac{3}{7}$.

$$12 = 11\dfrac{7}{7}$$

$$-8\dfrac{3}{7} = -8\dfrac{3}{7}$$

$$\boxed{3\dfrac{4}{7}}$$

> Borrow 1 from 12 and write it as $\dfrac{7}{7}$. Now subtract the fractions and subtract the whole numbers.

Your turn:

6. Subtract: $6 - 2\dfrac{4}{9}$.

Complete this example:

7. Off the coast of Massachusetts, the legal lobster size increased from $3\dfrac{13}{32}$ inches to $3\dfrac{1}{2}$ inches. How much of an increase was this?

To find the amount of increase you must subtract. Show your work below:

Your turn:

8. If Tucson's average annual rainfall is $11\dfrac{3}{4}$ inches and Yuma's is $3\dfrac{3}{5}$ inches, how much more rain, on average, does Tucson get than Yuma?

	Answer	Text Ref	Video Ref		Answer	Text Ref	Video Ref
1	Exact: $7\dfrac{17}{24}$ Approx: 7	Ex 1, p. 204		**5**	$3\dfrac{4}{7}$	Ex 6, p. 207	
2	Exact: $13\dfrac{11}{14}$ Approx: 14		Sec 3.4, 1/4	**6**	$3\dfrac{5}{9}$		Sec 3.4, 3/4
3	$5\dfrac{1}{15}$	Ex 2, p. 205		**7**	$\dfrac{3}{32}$ in.	Ex 8, p. 208	
4	$7\dfrac{5}{24}$		Sec 3.4, 2/4	**8**	$7\dfrac{13}{20}$ in.		Sec 3.4, 4/4

☐   **Next, insert your homework.** Make sure you attempt all exercises asked of you and show all work, as in the exercises above. Check your answers if possible. Clearly mark any exercises you were unable to correctly complete so that you may ask questions later. DO NOT ERASE YOUR INCORRECT WORK. THIS IS HOW WE UNDERSTAND AND EXPLAIN TO YOU YOUR ERRORS.

Section 3.5 Order, Exponents, and the Order of Operations

Before Class:

☐   Read the objectives on page 215.

☐   Read the **Helpful Hint** box on page 216.

☐   Complete the exercises:

     1.   Insert < or > to form a true sentence.

$$\frac{5}{7} \qquad \frac{2}{7}$$

     2.   Simplify: $\dfrac{4}{5} \cdot \dfrac{4}{5}$ 3.   Simplify: $\dfrac{6}{5} \div \dfrac{12}{35}$ 4.   Simplify: $\dfrac{6}{5} + \dfrac{13}{35}$

During Class:

☐   **Write your class notes.** Neatly write down **all** examples shown as well as key terms or phrases with definitions. If not applicable or if you were absent, watch the Lecture Series (DVD) for this section and do the same (write down the examples shown as well as key terms or phrases). Insert more paper as needed.

Class Notes/Examples	Your Notes

Answers: **1)** >   **2)** $\dfrac{16}{25}$ **3)** $\dfrac{7}{2}$ **4)** $\dfrac{11}{7}$

Section 3.5 Order, Exponents, and the Order of Operations

Class Notes (continued)	**Your Notes**

(Insert additional paper as needed.)

Section 3.5 Order, Exponents, and the Order of Operations

Practice:

☐   Complete the Vocabulary and Readiness Check on page 219.

☐   Complete any incomplete exercises below. Check and correct your work using the answers and references at the end of this section.

Review this example:

1. Insert < or > to form a true sentence.

$\dfrac{9}{10}$ $\dfrac{11}{12}$ The LCD for these fractions is 60.

Write each fraction as an equivalent fraction with a denominator of 60.

$$\frac{9}{10} = \frac{9 \cdot 6}{10 \cdot 6} = \frac{54}{60} \qquad \frac{11}{12} = \frac{11 \cdot 5}{12 \cdot 5} = \frac{55}{60}$$

Since 54 < 55, then

$$\frac{54}{60} < \frac{55}{60} \quad \text{or} \quad \boxed{\frac{9}{10} < \frac{11}{12}}$$

Your turn:

2. Insert < or > to form a true sentence.

$\dfrac{3}{5}$ $\dfrac{9}{14}$

Complete this example:

3. a. Evaluate the expression: $\left(\dfrac{3}{5}\right)^3$

$$\left(\frac{3}{5}\right)^3 = \left(\frac{3}{5}\right)\left(\frac{3}{5}\right)\left(\frac{3}{5}\right) = \boxed{\underline{\qquad}}$$

b. Evaluate the expression: $\left(\dfrac{1}{6}\right)^2 \cdot \left(\dfrac{3}{4}\right)^3$

$$\left(\frac{1}{6}\right)^2 \cdot \left(\frac{3}{4}\right)^3 =$$

$$\left(\frac{1}{6} \cdot \frac{1}{6}\right) \cdot \left(\frac{3}{4} \cdot \frac{3}{4} \cdot \frac{3}{4}\right) = \frac{1 \cdot 1 \cdot \cancel{3} \cdot \cancel{3} \cdot 3}{2 \cdot \cancel{3} \cdot 2 \cdot \cancel{3} \cdot 4 \cdot 4 \cdot 4} = \boxed{\underline{\qquad}}$$

Your turn:

4. Evaluate the expression: $\left(\dfrac{2}{5}\right)^3$

Section 3.5 Order, Exponents, and the Order of Operations

Complete this example:

5. Simplify: $\dfrac{1}{5} \div \dfrac{2}{3} \cdot \dfrac{4}{5}$ | Multiply or divide in order from left to right. |

$\dfrac{1}{5} \div \dfrac{2}{3} \cdot \dfrac{4}{5} = \dfrac{1}{5} \cdot \dfrac{3}{2} \cdot \dfrac{4}{5}$ | To divide, multiply by the reciprocal. |

$\dfrac{1}{5} \div \dfrac{2}{3} \cdot \dfrac{4}{5} = \dfrac{3}{10} \cdot \dfrac{4}{5} = \bigcirc \underline{\qquad}$

The answer in reduced form : $\bigcirc \underline{\qquad}$

Your turn:

6. Simplify: $\dfrac{1}{5} + \dfrac{1}{3} \cdot \dfrac{1}{4}$

Complete this example:

7. Simplify: $\left(\dfrac{2}{3}\right)^2 \div \left(\dfrac{8}{27} + \dfrac{2}{3}\right)$

$\left(\dfrac{2}{3}\right)^2 \div \left(\dfrac{8}{27} + \dfrac{2 \cdot 9}{3 \cdot 9}\right) = \left(\dfrac{2}{3}\right)^2 \div \dfrac{26}{27}$ | LCD is 27. |

Square $\dfrac{2}{3}$ and multiply by the reciprocal.

$\left(\dfrac{2}{3}\right)^2 \div \dfrac{26}{27} = \dfrac{4}{9} \div \dfrac{26}{27} = \dfrac{\cancel{4}^{\,2}}{\cancel{9}_{\,1}} \cdot \dfrac{\cancel{27}^{\,3}}{\cancel{26}_{\,13}} = \bigcirc \underline{\qquad}$

Your turn:

8. Simplify: $\left(\dfrac{2}{3} - \dfrac{5}{9}\right)^2$

	Answer	Text Ref	Video Ref		Answer	Text Ref	Video Ref
1	$\dfrac{9}{10} < \dfrac{11}{12}$	Ex 2, p. 216		**5**	$\dfrac{12}{50} = \dfrac{6}{25}$	Ex 10, p. 218	
2	$\dfrac{3}{5} < \dfrac{9}{14}$		Sec 3.5, 2/6	**6**	$\dfrac{17}{60}$		Sec 3.5, 5/6
3	a. $\dfrac{27}{125}$ b. $\dfrac{3}{256}$	Ex 4 − 5, p. 216		**7**	$\dfrac{6}{13}$	Ex 11, p. 218	
4	$\dfrac{8}{125}$		Sec 3.5, 3/6	**8**	$\dfrac{1}{81}$		Sec 3.5, 6/6

☐   **Next, insert your homework.** Make sure you attempt all exercises asked of you and show all work, as in the exercises above. Check your answers if possible. Clearly mark any exercises you were unable to correctly complete so that you may ask questions later. DO NOT ERASE YOUR INCORRECT WORK. THIS IS HOW WE UNDERSTAND AND EXPLAIN TO YOU YOUR ERRORS.

Preparing for the Chapter 3 Test

Before Class:

☐ Read the objectives on page 223.

☐ Read the Read the **Helpful Hint** box on page 224.

☐ Read the **Problem-Solving Steps** on page 223.

☐ Complete the exercises:

1. If the radius of a circle is 12 feet, what is the diameter of the circle?

2. A page in a book measures 27.5 cm by 20.5 cm. Find its area.

3. Find the volume of a box in the shape of a cube that is 5 feet on each side.

During Class:

☐ **Write your class notes.** Neatly write down **all** examples shown as well as key terms or phrases with definitions. If not applicable or if you were absent, watch the Lecture Series (DVD) for this section and do the same (write down the examples shown as well as key terms or phrases). Insert more paper as needed.

Class Notes/Examples	Your Notes

Answers: **1)** 24 *ft*   **2)** 563.75 *sq cm*   **3)** 125 *cu ft*

Section 3.5 Order, Exponents, and the Order of Operations

Class Notes (continued)	**Your Notes**

(Insert additional paper as needed.)

Preparing for the Chapter 3 Test

Practice:

☐   Complete any incomplete exercises below. Check and correct your work using the answers and references at the end of this section.

Review this example:

1. A Sony camcorder measures 5 *inches* by $2\frac{1}{2}$ *inches* by $1\frac{3}{4}$ *inches* . Find the volume of the camcorder box.

UNDERSTAND & TRANSLATE

To find the volume of a box multiply its length, width and height.

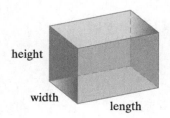

Volume = *length* × *width* × *height*

Volume = $5\ in. \times 2\frac{1}{2}\ in. \times 1\frac{3}{4}\ in.$

SOLVE

Volume = $\frac{5}{1} \cdot \frac{5}{2} \cdot \frac{7}{4} = \frac{175}{8} = \left(21\frac{7}{8}\ cubic\ inches\right)$

INTERPRET

The volume of the camcorder is $\left(21\frac{7}{8}\ cubic\ inches\right)$

Your turn:

2. Early cell phones were large and heavy. One early model measured approximately 8 *inches* by $2\frac{1}{2}$ *inches* by $2\frac{1}{2}$ *inches* . Find the volume of a box with those dimensions.

Section 3.5 Order, Exponents, and the Order of Operations

Review this example:

3. Find the total length of the given diagram.

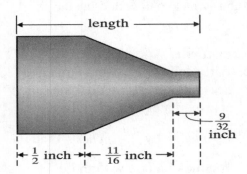

UNDERSTAND & TRANSLATE

The word total length tells us to add the lengths.

Total length = $\frac{1}{2}$ *in.* + $\frac{11}{16}$ *in.* + $\frac{9}{32}$ *in.*

SOLVE & INTERPRET

$$\frac{1}{2} + \frac{11}{16} + \frac{9}{32} = \frac{1 \cdot 16}{2 \cdot 16} + \frac{11 \cdot 2}{16 \cdot 2} + \frac{9}{32} = \frac{16}{32} + \frac{22}{32} + \frac{9}{32}$$

Total Length = $\frac{47}{32} = 1\frac{15}{32}$ *inches*

Your turn:

4. Suppose that the cross section of a piece of pipe looks like the diagram shown. Find the total outer diameter.

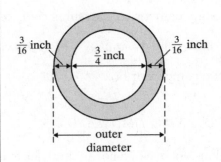

	Answer	Text Ref	Video Ref		Answer	Text Ref	Video Ref
1	$21\frac{7}{8}$ *cu in.*	Ex 1, p. 224		3	$\frac{47}{32} = 1\frac{15}{32}$ *in.*	Ex 2, p. 225	
2	50 *cu in.*		Sec 3.6, p. 230 #34 (not a video)	4	$\frac{9}{8} = 1\frac{1}{8}$ *in.*		Sec 3.6, 2/2

☐ **Next, insert your homework.** Make sure you attempt all exercises asked of you and show all work, as in the exercises above. Check your answers if possible. Clearly mark any exercises you were unable to correctly complete so that you may ask questions later. DO NOT ERASE YOUR INCORRECT WORK. THIS IS HOW WE UNDERSTAND AND EXPLAIN TO YOU YOUR ERRORS.

Preparing for the Chapter 3 Test

Start preparing for your Chapter 3 Test as soon as possible. Pay careful attention to any instructor discussion about this test, especially discussion on what sections you will be responsible for, etc.

☐ Work the Chapter 3 Vocabulary Check on page 234.

☐ Read both columns (Definitions and Concepts, and Examples) of the Chapter 3 Highlights starting on page 234.

☐ Read your Class Notes/Examples for each section covered on your Chapter 3 Test. Look for any unresolved questions you may have.

☐ Complete as many of the Chapter 3 Review exercises as possible (starting on page 237). Remember, the odd answers are in the back of your text.

☐ **Most important:** Place yourself in "test" conditions (see below) and work the Chapter 3 Test (pages 242-243) as a practice test the day before your actual test. To honestly assess how you are doing, try the following:

- Work on a few blank sheets of paper.
- Give yourself the same amount of time you will be given for your actual test.
- Complete this Chapter 3 Practice Test without using your notes or your text.
- If you have any time left after completing this practice test, check your work and try to find any errors on your own.
- Once done, use the back of your book to check ALL answers.
- Try to correct any errors on your own.
- Use the Chapter Test Prep Video (CTPV) to correct any errors you were unable to correct on your own. You can find these videos in the Interactive DVD Lecture Series, in MyMathLab, and on YouTube. Search MartinGayDevMath and click "Channels."

I wish you the best of luck….Elayn Martin-Gay

Section 4.1 Introduction to Decimals

Before Class:

☐ Read the objectives on page 248.

☐ Read the **Helpful Hint** boxes on pages 248 and 250.

☐ Complete the exercises:

1. On the place value chart, the place values to the **left** of the decimal point end in "_____".

   Place values to the **right** of the decimal point end in "_____".

2. Determine the place value for the digit **7** in each number.

   a. 70   b. 700   c. 0.7   d. 0.07

During Class:

☐ **Write your class notes.** Neatly write down **all** examples shown as well as key terms or phrases with definitions. If not applicable or if you were absent, watch the Lecture Series (DVD) for this section and do the same (write down the examples shown as well as key terms or phrases). Insert more paper as needed.

Class Notes/Examples	**Your Notes**

Answers: **1)** s ; ths   **2)** a. tens   b. hundreds   c. tenths   d. hundredths

Section 4.1 Introduction to Decimals

Class Notes (continued)	**Your Notes**

(Insert additional paper as needed.)

Section 4.1 Introduction to Decimals

Practice:

☐   Complete the Vocabulary and Readiness Check on page 252.

☐   Next, complete any incomplete exercises below. Check and correct your work using the answers and references at the end of this section.

Review this example:
1. Write each decimal number in words.

  a. 1.3

  b. 19.5023

  a. one and three tenths

  b. nineteen and five thousand, twenty-three ten-thousandths

Your turn:
2. Write each decimal number in words.

  a. 16.23

  b. 167.009

Complete this example:
3. Write each decimal number in standard form.

  a. forty-eight and twenty-six hundredths

  b. six and ninety-five thousandths

  a. 48._____   b. 6._____

Your turn:
4. Write each decimal number in standard form.

  a. nine and eight hundredths

  b. forty-six ten-thousandths

Complete this example:
5. Write as a fraction or mixed number. Write your answer in simplest form.

  a. 0.125   b. 23.5

  a. $0.125 = \dfrac{125}{1000} = \dfrac{125}{125 \cdot 8} = \bigcirc$

  b. $23.5 = 23\dfrac{5}{10} = 23\dfrac{5}{2 \cdot 5} = \bigcirc$

Your turn:
6. Write as a fraction or mixed number. Write your answer in simplest form.

  a. 0.27

  b. 7.008

85

Section 4.1 Introduction to Decimals

Review this example:

7. Write each fraction as a decimal.

a. $\dfrac{87}{10}$ b. $\dfrac{18}{1000}$

a. $\dfrac{87}{10} = 8.7$ b. $\dfrac{18}{1000} = 0.018$

↑   ↑   ↑   ↑

1 zero   1 decimal place   3 zeros   3 decimal places

Your turn:

8. Write each fraction as a decimal.

a. $\dfrac{6}{10}$ b. $\dfrac{45}{100}$

	Answer	Text Ref	Video Ref		Answer	Text Ref	Video Ref
1	a. one and three tenths b. nineteen and five thousand, twenty-three ten-thousandths	Ex 1–3, p. 249		5	a. $\dfrac{1}{8}$ b. $23\dfrac{1}{2}$	Ex 10–11, p. 251	
2	a. sixteen and twenty-three hundredths b. one hundred sixty-seven and nine thousandths		Sec 4.1, 1-2/9	6	a. $\dfrac{27}{100}$ b. $7\dfrac{1}{125}$		Sec 4.1, 5-6/9
3	a. 48.26 b. 6.095	Ex 6–7, p. 250		7	a. 8.7 b. 0.018	Ex 14–15 p. 251	
4	a. 9.08 b. 0.0046		Sec 4.1, 3-4/9	8	a. 0.6 b. 0.45		Sec 4.1, 7-8/9

☐   **Next, insert your homework.** Make sure you attempt all exercises asked of you and show all work, as in the exercises above. Check your answers if possible. Clearly mark any exercises you were unable to correctly complete so that you may ask questions later. DO NOT ERASE YOUR INCORRECT WORK. THIS IS HOW WE UNDERSTAND AND EXPLAIN TO YOU YOUR ERRORS.

Section 4.2 Order and Rounding

Before Class:

☐   Read the objectives on page 256.

☐   Read the **Helpful Hint** boxes on pages 256 and 257.

☐   Complete the exercises:

1.   For any decimal, writing 0's after the last digit to the right of the decimal does/does not change the number.

7.4 = _____ = _____

2.   Round 47,261 to the indicated place value.

a. tens   b. hundreds   c. thousands   d. ten-thousands

During Class:

☐   **Write your class notes.** Neatly write down **all** examples shown as well as key terms or phrases with definitions. If not applicable or if you were absent, watch the Lecture Series (DVD) for this section and do the same (write down the examples shown as well as key terms or phrases). Insert more paper as needed.

Class Notes/Examples	Your Notes

Answers: **1)**   does not; 7.4 = 7.40 = 7.400   **2)** a. 47,260   b. 47,300   c. 47,000   d. 50,000

Section 4.2 Order and Rounding

Class Notes (continued)	Your Notes

(Insert additional paper as needed.)

Section 4.2 Order and Rounding

Practice:

☐   Complete the Vocabulary and Readiness Check on page 260.

☐   Next, complete any incomplete exercises below. Check and correct your work using the answers and references at the end of this section.

Review this example:	**Your turn:**
1. Insert <, >, or = to form a true statement.	**2.** Insert <, >, or = to form a true statement.
0.378 0.368	
	0.57 0.54
0.378 0.368 Tenths places are the same.	
0.378 0.368 Hundredths places are different.	
Since 7 > 6, then 0.378 > 0.368.	

Complete this example:	**Your turn:**
3. Insert <, >, or = to form a true statement.	**4.** Insert <, >, or = to form a true statement.
0.052 0.236	
	167.908 167.980
0.052 0.236 Notice the digits in the tenths places are _____ .	
Since 0 ☐ 2, then 0.052 ☐ 0.236	

Review this example:	**Your turn:**
5. Round each decimal to the given value.	**6.** Round each decimal to the given value.
736.2359 , to the nearest tenth	
	0.234 , to the nearest hundredth
Step 1. Locate the digit to the right of the tenths place.	
736.2359 3 is the digit to the right	
Step 2. Since the digit to the right is less than 5, we delete it and all digits to its right.	
736.2359 rounds to 736.2.	

89

Section 4.2 Order and Rounding

Review this example:	**Your turn:**
7. Round 736.2359, to the nearest hundredth.	**8.** Round $98,207.23$, to the nearest ten.

Step 1. Locate the digit to the right of the hundredths place.

736.23<u>5</u>9   5 is the digit to the right

Step 2. Since the digit to the right is 5, we add 1 to the digit in the hundredths place and delete all digits to the right of the hundredths place.

736.2359 rounds to $\boxed{736.24.}$

Complete this example:	**Your turn:**
9. Round $\$3.1779$, to the nearest cent.	**10.** Round $\$26.95$, to the nearest dollar.

$\$3.17\underline{7}9$ Since 7 is the digit to the right and greater than 5, we add 1 to the hundredths place digit and delete all digits to the right of the hundredths place digit.

$\$3.1779$ rounds to   ⬭

	Answer	Text Ref	Video Ref		Answer	Text Ref	Video Ref
1	>	Ex 1, p. 257		6	0.23		Sec 4.2, 3/7
2	>		Sec 4.2, 1/7	7	736.24	Ex 6, p. 258	
3	different ; <	Ex 2, p. 257		8	98,210		Sec 4.2, 4/7
4	<		Sec 4.2, 2/7	9	$3.18	Ex 7, p. 258	
5	736.2	Ex 5, p. 258		10	$27		Sec 4.2, 6/7

☐   **Next, insert your homework.** Make sure you attempt all exercises asked of you and show all work, as in the exercises above. Check your answers if possible. Clearly mark any exercises you were unable to correctly complete so that you may ask questions later. DO NOT ERASE YOUR INCORRECT WORK. THIS IS HOW WE UNDERSTAND AND EXPLAIN TO YOU YOUR ERRORS.

Section 4.3 Adding and Subtracting Decimals

Before Class:

☐   Read the objectives on page 264.

☐   Read the **Helpful Hint** boxes on pages 264, 265, and 266.

☐   Complete the exercises:

1.   2849.1738 rounded to the nearest hundred is _____.

2.   146.059 rounded to the nearest ten is _____.

3.   2849.1738 rounded to the nearest hundredth is _____.

During Class:

☐   **Write your class notes.** Neatly write down **all** examples shown as well as key terms or phrases with definitions. If not applicable or if you were absent, watch the Lecture Series (DVD) for this section and do the same (write down the examples shown as well as key terms or phrases). Insert more paper as needed.

Class Notes/Examples	Your Notes

Answers: **1)** 2800   **2)** 150   **3)** 2849.17

Section 4.3 Adding and Subtracting Decimals

Class Notes (continued)	**Your Notes**

(Insert additional paper as needed.)

Section 4.3 Adding and Subtracting Decimals

Practice:

☐   Complete the Vocabulary and Readiness Check on page 269.

☐   Next, complete any incomplete exercises below. Check and correct your work using the answers and references at the end of this section.

Review this example:
1. Add: $763.7651 + 22.001 + 43.89$

Line up the decimal points.

$$\begin{array}{r} \overset{1\ \ \ 1\ \ \ 1}{763.7651} \\ 22.0010 \\ +\ \ \ 43.8900 \\ \hline 829.6561 \end{array}$$

22.0010   Insert one zero.
43.8900   Insert two zeros.
829.6561   Add.

Your turn:
2. Add: $24.6 + 2.39 + 0.0678$

Review this example:
3. Subtract and check: $85 - 17.31$

Note: The decimal point in the whole number 85 is located after the last digit.

$$\begin{array}{r} \overset{14\ \ \ 9}{7\ \ \overset{}{\cancel{4}}\ \overset{10}{\cancel{10}}\ \overset{10}{\cancel{10}}} \\ \cancel{8}\cancel{5}.\cancel{0}\cancel{0} \\ -\ 1\ 7.3\ 1 \\ \hline 6\ 7.6\ 9 \end{array}$$

Insert two zeros.

Check:
$$\begin{array}{r} 67.69 \\ +\ 17.31 \\ \hline 85.00 \end{array}$$

Your turn:
4. Subtract and check: $18 - 2.7$

Review this example:
5. Subtract: $11.01 - 0.862$. Then estimate to determine if answer is reasonable.

Exact: Estimate:

Round both numbers to the nearest one.

$$\begin{array}{r} \overset{0\ \ \ 9\ \ 10\ \ 10}{1\overset{14}{\cancel{1}}.\cancel{0}\overset{9}{\cancel{1}}\cancel{0}} \\ -\ 0.8\ 6\ 2 \\ \hline 1\ 0.1\ 4\ 8 \end{array}$$

rounds to   11
rounds to -1
                  ——
                   10

Your turn:
6. Subtract: $1000 - 123.4$. Then estimate to determine if your answer is reasonable.

Section 4.3 Adding and Subtracting Decimals

Review this example:

7. Find the total monthly cost of owning and operating a certain automobile given the expenses shown.

Monthly car payment:	$256.63
Monthly insurance cost:	$47.52
Avg. gasoline bill per month:	$95.33

UNDERSTAND:
The phrase "total monthly cost" tells us to add.

TRANSLATE and *SOLVE*:

total monthly   car   insurance   gas
  cost   = payment + cost   + bill
   ↓   ↓   ↓   ↓

total monthly = $256.63 + $47.52 + $95.33

$$\begin{array}{r} \overset{1\ 1\ 1}{25}6.63 \\ 47.52 \\ +\ 95.33 \\ \hline \$399.48 \end{array}$$

INTERPRET:
The total monthly cost is $399.48.

Your turn:

8. A landscape architect is planning a border for a flower garden shaped like a triangle. The sides of the garden measure 12.4 feet, 29.34 feet, and 25.7 feet. Find the amount of border material needed.

	Answer	Text Ref	Video Ref		Answer		Text Ref	Video Ref
1	829.6561	Ex 2, p. 265		5	Exact: 10.148		Ex 7b, p. 266	
2	27.0578		Sec 4.3, 2/6	6	Exact: 876.6	Est: 900		Sec 4.3, 5/6
3	67.69	Ex 6, p. 266		7	$399.48		Ex 8, p. 267	
4	15.3		Sec 4.3, 3/6	8	67.44 *ft*			Sec 4.3, 6/6

☐   **Next, insert your homework.** Make sure you attempt all exercises asked of you and show all work, as in the exercises above. Check your answers if possible. Clearly mark any exercises you were unable to correctly complete so that you may ask questions later. DO NOT ERASE YOUR INCORRECT WORK. THIS IS HOW WE UNDERSTAND AND EXPLAIN TO YOU YOUR ERRORS.

Section 4.4 Multiplying Decimals and Circumference of a Circle

Before Class:

☐   Read the objectives on page 275.

☐   Read the Rules and Definition boxes on pages 275, 277, and 278.

☐   Complete the exercises:

1. Find the perimeter of a square with sides of length 45.2 cm.

2. Find the perimeter of the triangle pictured.

4.2 in.   5.78 in.

7.8 in.

During Class:

☐   **Write your class notes.** Neatly write down **all** examples shown as well as key terms or phrases with definitions. If not applicable or if you were absent, watch the Lecture Series (DVD) for this section and do the same (write down the examples shown as well as key terms or phrases). Insert more paper as needed.

Class Notes/Examples	Your Notes

Answers: **1)** 180.8 cm   **2)** 17.78 in.

Section 4.4 Multiplying Decimals and Circumference of a Circle

Class Notes (continued)	**Your Notes**

(Insert additional paper as needed.)

Section 4.4 Multiplying Decimals and Circumference of a Circle

Practice:

☐   Complete the Vocabulary and Readiness Check on page 280.

☐   Next, complete any incomplete exercises below. Check and correct your work using the answers and references at the end of this section.

Review this example:

1. Multiply: 23.6×0.78

Step 1: Multiply the decimals as though they are whole numbers.

Step 2: The decimal place of the product is equal to the sum of the number of decimal places in the factors.

$$
\begin{array}{r}
23.6 \\
\times\ 0.78 \\
\hline
1888 \\
16520 \\
\hline
18.408 \\
\end{array}
$$

23.6 ← **1** decimal place
× 0.78 ← **2** decimal places

Since **1 + 2 = 3**, insert the decimal point in the product so that there are **3** decimal places.

Your turn:

2. Multiply: 1.2×0.5

Review this example:

3. Multiply: 28.06×1.95. Then estimate to determine if your answer is reasonable.

$$
\begin{array}{r}
28.06 \\
\times\ 1.95 \\
\hline
14030 \\
252540 \\
280600 \\
\hline
54.7170 \\
\end{array}
$$

Estimate 1:   Estimate 2:
(Rounded to ones)   (Rounded to tens)

$$
\begin{array}{r}
28 \\
\times 2 \\
\hline
56 \\
\end{array}
\qquad
\begin{array}{r}
30 \\
\times 2 \\
\hline
60 \\
\end{array}
$$

The answer is reasonable.

Your turn:

4. Multiply: 1.0047×8.2. Then estimate to determine if your answer is reasonable.

Review this example:

5. Multiply: 23.702×100

Is the decimal being multiplied by a power of 10 such as 10, 100, 1000, …? (Yes) No

How many zeros are in the power of 10? (2)

Move the decimal point **2** places to the **right**.
$23.702 \times 100 = (2370.2)$

Your turn:

6. Multiply: 7.093×100

97

Section 4.4 Multiplying Decimals and Circumference of a Circle

| **Review this example:** | **Your turn:** |

Review this example:

7. Multiply: $76,805 \times 0.01$

Is the decimal being multiplied by a power of 10 such as 0.1, 0.01, 0.001, …? (Yes) No

Move the decimal point **left** the same number of places as there are decimal places in the power of 10. (0.01 has two decimal places.)

$76,805 \times 0.01 = \boxed{768.05}$

Your turn:

8. Multiply: $37.62 \times 0.001 =$

Complete this example:

9. Find the circumference of the circle. Use $\pi \approx 3.14$ to approximate the circumference.

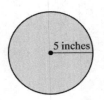

5 inches

$C = 2\pi r = 2 \cdot \pi \cdot 5 = \boxed{10\pi \, in.}$ Exact answer

$\approx 10(3.14)$ in.

$\approx \boxed{}$ Approximate answer

Your turn:

10. Find the circumference of a circle whose diameter is 4 m. Then, use $\pi \approx 3.14$ to approximate the circumference.

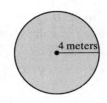

4 meters

	Answer	Text Ref	Video Ref		Answer	Text Ref	Video Ref
1	18.408	Ex 1, p. 275		6	709.3		Sec 4.4, 3/6
2	0.6		Sec 4.4, 1/6	7	768.05	Ex 9, p. 277	
3	Exact: 54.7170	Ex 4, p. 276		8	0.03762		Sec 4.4, 4/6
4	Exact: 8.23854 Est: 8		Sec 4.4, 2/6	9	exact; approx 10π in. ≈ 31.4 in.	Ex 12, p. 278	
5	2370.2	Ex 6, p. 277		10	exact; approx 8π m ≈ 25.12 m		Sec 4.4, 5/6

☐ **Next, insert your homework.** Make sure you attempt all exercises asked of you and show all work, as in the exercises above. Check your answers if possible. Clearly mark any exercises you were unable to correctly complete so that you may ask questions later. DO NOT ERASE YOUR INCORRECT WORK. THIS IS HOW WE UNDERSTAND AND EXPLAIN TO YOU YOUR ERRORS.

Section 4.5 Dividing Decimals and Order of Operations

Before Class:

☐   Read the objectives on page 287.

☐   Read the steps for **Dividing by a Decimal** on page 288, the steps for **Dividing Decimals by Powers of 10** on page 290, and the steps for the **Order of Operations** on page 291.

☐   Complete the exercises:

1. In the division problem $2.5\overline{\smash{)}2.88}$, move the decimal point in both numbers _____ place(s) to the _____.

2. In the division problem $0.25\overline{\smash{)}2.88}$, move the decimal point in both numbers _____ places(s) to the _____.

3. Round 24.683 to the nearest tenth. _____

4. Divide 432.7 by 1000 . Move the decimal point _____ places to the _____ . What number do you get? _____

During Class:

☐   **Write your class notes.** Neatly write down **all** examples shown as well as key terms or phrases with definitions. If not applicable or if you were absent, watch the Lecture Series (DVD) for this section and do the same (write down the examples shown as well as key terms or phrases). Insert more paper as needed.

Class Notes/Examples	Your Notes

Answers: **1)** 1, right   **2)** 2, right   **3)** 24.7   **4)** three, left, 0.4327

Section 4.5 Dividing Decimals and Order of Operations

Class Notes (continued)	**Your Notes**

(Insert additional paper as needed.)

Section 4.5 Dividing Decimals and Order of Operations

Practice:

☐   Complete the Vocabulary and Readiness Check on page 293.

☐   Next, complete any incomplete exercises below. Check and correct your work using the answers and references at the end of this section.

Review this example:

1. Divide: $10.764 \div 2.3$

Move the decimal points in the divisor and the dividend one place to the right, so that 2.3 is a whole number.

$2.3 \overline{)10.764} \quad \Rightarrow \quad 23 \overline{)107.64}$

$10.764 \div 2.3 = 4.68$

$$
\begin{array}{r}
4.68 \\
23 \overline{)107.64} \\
-92 \\
\hline
15\,6 \\
-13\,8 \\
\hline
1\,84 \\
-1\,84 \\
\hline
0
\end{array}
$$

Your turn:

2. Divide: $4.756 \div 0.82$

$0.82 \overline{)4.756} \quad \Rightarrow$

Review this example:

3. Divide: $17.5 \div 0.48$. Round the quotient to the nearest hundredth.

$0.48 \overline{)17.5} \quad \Rightarrow \quad 48 \overline{)1750.000}$

$$
\begin{array}{r}
36.458 \quad \approx 36.46 \\
48 \overline{)1750.000} \\
-144 \\
\hline
310 \\
-288 \\
\hline
22\,0 \\
-19\,2 \\
\hline
2\,80 \\
-2\,40 \\
\hline
400 \\
-384 \\
\hline
16
\end{array}
$$

Your turn:

4. Divide: $0.549 \div 0.023$. Round the quotient to the nearest tenth.

101

Section 4.5 Dividing Decimals and Order of Operations

Review this example:

5. Divide: $\dfrac{786.1}{1000}$

Is the decimal being divided by a power of 10 such as 10, 100, 1000, …? (Yes) No

How many zeros are in the power of 10? ③
Move the decimal point 3 places to the **left**.

$\dfrac{786.1}{1000} = \boxed{0.7861}$

Your turn:

6. Divide: $\dfrac{54.982}{100}$

Complete this example:

7. Simplify the expression: $0.5(8.6 - 1.2)$

Following the order of operations, we simplify inside the parentheses first.

$0.5(8.6 - 1.2) = 0.5(7.4)$ Subtract.
$\boxed{= 3.7}$ Multiply.

Your turn:

8. Simplify the expression:

$0.7(6 - 2.5)$

	Answer	Text Ref	Video Ref		Answer	Text Ref	Video Ref
1	4.68	Ex 4, p. 289		5	0.7861	Ex 8, p. 290	
2	5.8		Sec 4.5, 2/9	6	0.54982		Sec 4.5, 5/9
3	≈ 36.46	Ex 6, p. 289		7	3.7	Ex 12, p. 292	
4	23.87		Sec 4.5, 3/9	8	2.45		Sec 4.5, 8/9

☐   **Next, insert your homework.** Make sure you attempt all exercises asked of you and show all work, as in the exercises above. Check your answers if possible. Clearly mark any exercises you were unable to correctly complete so that you may ask questions later. DO NOT ERASE YOUR INCORRECT WORK. THIS IS HOW WE UNDERSTAND AND EXPLAIN TO YOU YOUR ERRORS.

Section 4.6 Fractions and Decimals

Before Class:

☐   Read the objectives on page 298.

☐   Complete the exercises:

1. Write $\dfrac{4}{5}$ as an equivalent fraction with a denominator of 10 .

$$\frac{4}{5} = \frac{4}{5} \cdot \frac{2}{2} = \frac{}{10}$$

2. Write equivalent fractions with a denominator of 100 for the following:

a. $\dfrac{7}{50}$

b. $\dfrac{3}{25}$

During Class:

☐   **Write your class notes.**   Neatly write down **all** examples shown as well as key terms or phrases with definitions. If not applicable or if you were absent, watch the Lecture Series (DVD) for this section and do the same (write down the examples shown as well as key terms or phrases). Insert more paper as needed.

Class Notes/Examples	Your Notes

Answers: **1)** $\dfrac{8}{10}$ **2)** a. $\dfrac{14}{100}$ b. $\dfrac{12}{100}$

103

Section 4.6 Fractions and Decimals

Class Notes (continued)	**Your Notes**

(Insert additional paper as needed.)

Practice:

☐   Complete the Vocabulary and Readiness Check on page 302.

☐   Complete any incomplete exercises below. Check and correct your work using the answers and references at the end of this section.

Review this example:	**Your turn:**
1. Write $\dfrac{1}{4}$ as a decimal.	**2.** Write $\dfrac{3}{4}$ as a decimal.

$$\begin{array}{r} 0.25 \\ 4\overline{)1.00} \\ -8 \\ \hline 20 \\ -20 \\ \hline 0 \end{array}$$

Solution: $\left(\dfrac{1}{4}=0.25\right)$

Remainder is 2. Bring 0 down.

Review this example:	**Your turn:**
3. Write $\dfrac{2}{3}$ as a decimal.	**4.** Write $\dfrac{11}{12}$ as a decimal.

$$\begin{array}{r} 0.666\ldots \\ 3\overline{)2.000} \\ -1\,8 \\ \hline 20 \\ -18 \\ \hline 20 \\ -18 \\ \hline 2 \end{array}$$

Solution: $\left(\dfrac{2}{3}=0.666\ldots=0.\overline{6}\right)$

Remainder is 2. Bring 0 down.

Remainder is 2. Bring 0 down.

Remainder is 2. Place a bar over the 6 to show that it repeats.

105

Section 4.6 Fractions and Decimals

Review this example:

5. Insert <, >, or = between the numbers to form a true statement: $\dfrac{1}{8}$ 0.12

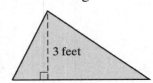

Original numbers	$\dfrac{1}{8}$	0.12
Decimals	0.125	0.120
Compare	0.125 > 0.12	

Thus, $\dfrac{1}{8} > 0.12$.

Your turn:

6. Insert <, >, or = between the numbers to form a true statement:

1.38 $\dfrac{18}{13}$

Review this example:

7. Find the area of the triangle shown. The base is 5.6 feet long.

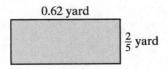

3 feet

$Area = \dfrac{1}{2} \cdot base \cdot height$

$= \dfrac{1}{2}(5.6)3 = 0.5(5.6)3 = \boxed{8.4 \text{ sq ft}}$

Your turn:

8. Find the area of the rectangle shown.

0.62 yard

$\dfrac{2}{5}$ yard

	Answer	Text Ref	Video Ref		Answer	Text Ref	Video Ref
1	0.25	Ex 1, p. 298		**5**	>	Ex 7, p. 300	
2	0.75		Sec 4.6, 1/4	**6**	<		Sec 4.6, 3/4
3	$0.\overline{6}$	Ex 2, p. 298		**7**	8.4 sq ft	Ex 10, p. 301	
4	$0.91\overline{6}$		Sec 4.6, 2/4	**8**	0.248 sq yd		Sec 4.6, 4/4

☐ **Next, insert your homework.** Make sure you attempt all exercises asked of you and show all work, as in the exercises above. Check your answers if possible. Clearly mark any exercises you were unable to correctly complete so that you may ask questions later. DO NOT ERASE YOUR INCORRECT WORK. THIS IS HOW WE UNDERSTAND AND EXPLAIN TO YOU YOUR ERRORS.

Preparing for the Chapter 4 Test

Start preparing for your Chapter 4 Test as soon as possible. Pay careful attention to any instructor discussion about this test, especially discussion on what sections you will be responsible for, etc.

☐   Work the Chapter 4 Vocabulary Check on page 307.

☐   Read both columns (Definitions and Concepts, and Examples) of the Chapter 4 Highlights starting on page 307.

☐   Read your Class Notes/Examples for each section covered on your Chapter 4 Test. Look for any unresolved questions you may have.

☐   Complete as many of the Chapter 4 Review exercises as possible (starting on page 310). Remember, the answers are in the back of your text.

☐   **Most important:** Place yourself in "test" conditions (see below) and work the Chapter 4 Test (pages 314 – 315) as a practice test the day before your actual test. To honestly assess how you are doing, try the following:
   - Work on a few blank sheets of paper.
   - Give yourself the same amount of time you will be given for your actual test.
   - Complete this Chapter 4 Practice Test without using your notes or your text.
   - If you have any time left after completing this practice test, check your work and try to find any errors on your own.
   - Once done, use the back of your book to check ALL answers.
   - Try to correct any errors on your own.
   - Use the Chapter Test Prep Video (CTPV) to correct any errors you were unable to correct on your own. You can find these videos in the Interactive DVD Lecture Series, in MyMathLab, and on YouTube. Search MartinGayDevMath and click "Channels."

I wish you the best of luck….Elayn Martin-Gay

Section 5.1 Ratio and Proportion

Before Class:

☐   Read the objectives on page 320.

☐   Read the Rules and Definition boxes on pages 320 and 323.

☐   Read the **Helpful Hint** boxes on page 320, 321, 322, 323 and 327.

☐   Complete the exercises:

1.   Write each fraction in simplest form: a. $\dfrac{15}{30}$ b. $\dfrac{36}{81}$ c. $\dfrac{40}{56}$

2.   Write the ratio of 3 to 4 using fraction notation and colon notation.

During Class:

☐   **Write your class notes.** Neatly write down **all** examples shown as well as key terms or phrases with definitions. If not applicable or if you were absent, watch the Lecture Series (DVD) for this section and do the same (write down the examples shown as well as key terms or phrases). Insert more paper as needed.

Class Notes/Examples	**Your Notes**

Answers: **1)** a. $\dfrac{1}{2}$ b. $\dfrac{4}{9}$ c. $\dfrac{5}{7}$ **2)** fraction notation is $\dfrac{3}{4}$, colon notation is $3:4$

Section 5.1 Ratio and Proportion

Class Notes (continued)	**Your Notes**

(Insert additional paper as needed.)

Practice:

☐   Complete the Vocabulary and Readiness Check on page 328.

☐   Next, complete any incomplete exercises below. Check and correct your work using the answers and references at the end of this section.

Review this example:	**Your turn:**
1. Write each ratio as a ratio of whole numbers using fractional notation. Write the fraction in simplest form.	**2.** Write each ratio as a ratio of whole numbers using fractional notation. Write the fraction in simplest form.

a. $15 to $10

$$\frac{\$15}{\$10} = \frac{15}{10} = \frac{3 \cdot \cancel{5}^{1}}{2 \cdot \cancel{5}_{1}} = \boxed{\frac{3}{2}}$$

a.   16 to 24

b.   2.6 to 3.1   Hint: Clear the decimals.

$$\frac{2.6}{3.1} = \frac{2.6}{3.1} \cdot 1 = \frac{2.6}{3.1} \cdot \frac{10}{10} = \frac{2.6 \cdot 10}{3.1 \cdot 10} = \boxed{\frac{26}{31}}$$

b.   7.7 to 10

Review this example:	**Your turn:**
3. Write each rate as a fraction in simplest form.	**4.** Write the rate as a fraction in simplest terms:

a. 360 miles on 16 gallons of gasoline

$$\frac{360 \ miles}{16 \ gallons} = \frac{45 \cdot \cancel{8} \ miles}{2 \cdot \cancel{8} \ gallons} = \boxed{\frac{45 \ miles}{2 \ gallons}}$$

a. 15 returns for 100 sales

b. $2160 for 12 weeks

$$\frac{2160 \ dollars}{12 \ weeks} = \frac{180 \cdot \cancel{12} \ dollars}{1 \cdot \cancel{12} \ weeks} = \boxed{\frac{180 \ dollars}{1 \ week}}$$

Write the rate as a unit rate.

b. 330 calories in a $3 - ounce$ serving

Explain why example 3b. is a unit rate when simplified? _____

Section 5.1 Ratio and Proportion

Review this example:

5. a.   Is $\dfrac{2}{3} = \dfrac{4}{6}$ a true proportion?

$2 \cdot 6 \overset{?}{=} 3 \cdot 4$ Find the cross products.

$12 = 12$ Since the cross products are equal, the proportion is (true.)

b.   Is $\dfrac{4.1}{7} = \dfrac{2.9}{5}$ a true proportion?

$(4.1) \cdot 5 \overset{?}{=} 7 \cdot (2.9)$ Find the cross products.

$20.5 \neq 20.3$ Since the cross products are **not** equal, the proportion is (false.)

Your turn:

6. Is $\dfrac{8}{6} = \dfrac{9}{7}$ a true proportion?

Review this example:

7. Find the value of n. $\dfrac{51}{34} = \dfrac{3}{n}$

$51 \cdot n = 34 \cdot 3$ Multiply.

$51n = 102$ Then, divide 102 by 51.

$\boxed{n = 2}$ Simplify.

Your turn:

8. Find the value of n. $\dfrac{n}{8} = \dfrac{50}{100}$

	Answer	Text Ref	Video Ref		Answer	Text Ref	Video Ref
1	a. $\dfrac{3}{2}$ b. $\dfrac{26}{31}$	Ex 2-3, p. 321		5	a. yes b. no	Ex 9-10, p. 323	
2	a. $\dfrac{2}{3}$ b. $\dfrac{77}{100}$		Sec 5.1, 2-3/13	6	It is false.		Sec 5.1, 7/13
3	a. $\dfrac{45\ miles}{2\ gallons}$ b. $\dfrac{\$180}{1\ week}$ denominator = one	Ex 7-8, p. 322		7	$n = 2$	Ex 11, p. 324	
4	a. $\dfrac{3\ returns}{20\ sales}$ b. $110\ cal\ /\ oz$		Sec 5.1, ?/13	8	$n = 4$		Sec 5.1, 9/13

☐   **Next, insert your homework.** Make sure you attempt all exercises asked of you and show all work, as in the exercises above. Check your answers if possible. Clearly mark any exercises you were unable to correctly complete so that you may ask questions later. DO NOT ERASE YOUR INCORRECT WORK. THIS IS HOW WE UNDERSTAND AND EXPLAIN TO YOU YOUR ERRORS.

Before Class:

☐   Read the objectives on page 336.

☐   Read the Rules and Definition boxes on pages 336, 337 and 338.

☐   Read the **Helpful Hint** boxes on page 337, 338 and 339.

☐   Complete the exercises:

1. Write the following fractions in simplest form:

   a. $\dfrac{5}{30}$ b. $\dfrac{42}{49}$ c. $\dfrac{10}{25}$

2. a. Percent means per _____.   b. 100% = _____.

3. a. $75\% = \dfrac{75}{?}$ b. $33\% = \dfrac{?}{100}$ c. $?\% = \dfrac{92.5}{100}$

During Class:

☐   **Write your class notes.** Neatly write down **all** examples shown as well as key terms or phrases with definitions. If not applicable or if you were absent, watch the Lecture Series (DVD) for this section and do the same (write down the examples shown as well as key terms or phrases). Insert more paper as needed.

Class Notes/Examples	Your Notes

Answers: **1)** a. $\dfrac{1}{6}$ b. $\dfrac{6}{7}$ c. $\dfrac{2}{5}$ **2)** a. one hundred b. 1   **3)** a. $\dfrac{75}{100}$ b. $\dfrac{33}{100}$ c. 92.5%

Section 5.2 Introduction to Percent

Class Notes (continued)	Your Notes

(Insert additional paper as needed.)

Practice:

☐   Complete the Vocabulary and Readiness Check on page 340.

☐   Next, complete any incomplete exercises below. Check and correct your work using the
         answers and references at the end of this section.

Review this example:	**Your turn:**
1. Silver (gray) has been the most popular color for cars in the United States the past nine years. Out of 100 people, 25 people drive silver cars. What percent of people drive silver cars? (*Source: U.S. News & World Report*)	**2.** In a survey of 100 college students, 96 use the Internet. What percent use the Internet?

Since 25 people out of 100 drive silver cars, the

fraction is $\left(\dfrac{25}{100} = 25\% \right)$

Review this example:	**Your turn:**
3. Write each percent as a decimal.	**4.** Write each percent as a decimal.
a. 23%	a. 6%
b. 0.74%	

Method 1: Replace the percent symbol with 0.01,
            then multiply.

a. $23\% = 23(0.01) = 0.23$ b. 2.8%

Method 2:

Percent→ | Remove the % symbol and move the decimal point 2 places to the **left**. | → Decimal

b. $0.74\% = 0.0074$

Section 5.2 Introduction to Percent

Complete this example:

5. Write each decimal as a percent.

    a. 0.65   b. 0.6

Method 1: Multiply by 100%.

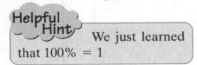
We just learned that $100\% = 1$

a. $0.65 = 0.65(100)\% = 65.\%$ or 65%

Method 2:

| Move the decimal point 2 places to the **right** and attach a % symbol. |

Percent ← | ← Decimal

b. $0.6 = 060.\% = \underline{\hspace{2cm}}\%$

Helpful Hint
A zero was inserted as a placeholder.

Your turn:

6. Write each decimal as a percent.

    a. 0.22

    b. 3.00

	Answer	Text Ref	Video Ref		Answer	Text Ref	Video Ref
1	25%	Ex 1, p. 336		**4**	a. 0.06 b. 0.028		Sec 5.2, 3&5/10
2	96%		Sec 5.2, 1/10	**5**	a. 65% b. 60%	Ex 8&11, p. 339	
3	a. 0.23 b. 0.0074	Ex 3&6, p. 338		**6**	a. 22% b. 300%		Sec 5.2, 7&9/10

☐   **Next, insert your homework.** Make sure you attempt all exercises asked of you and show all work, as in the exercises above. Check your answers if possible. Clearly mark any exercises you were unable to correctly complete so that you may ask questions later. DO NOT ERASE YOUR INCORRECT WORK. THIS IS HOW WE UNDERSTAND AND EXPLAIN TO YOU YOUR ERRORS.

Section 5.3 Percents and Fractions

Before Class:

☐　Read the objectives on page 344.

☐　Read the **Helpful Hint** boxes on pages 344 and 345.

☐　Read the Summary of Converting Percents, Decimals, and Fractions on page 346.

☐　Complete the exercises:

1.　Which of the following are correct?

　　a.　$6.5\% = 0.65$　　b.　$7.8\% = 0.078$　　c.　$120\% = 0.12$　　d.　$0.35\% = 0.0035$

2.　Which of the following are correct?

　　a.　$0.231 = 23.1\%$　　b.　$5.12 = 0.0512\%$　　c.　$3.2 = 320\%$　　d.　$0.0175 = 0.175\%$

During Class:

☐　**Write your class notes.**　Neatly write down **all** examples shown as well as key terms or phrases with definitions. If not applicable or if you were absent, watch the Lecture Series (DVD) for this section and do the same (write down the examples shown as well as key terms or phrases). Insert more paper as needed.

Class Notes/Examples	**Your Notes**

Answers: **1)**　b, d　　　　**2)**　a, c

Section 5.3 Percents and Fractions

Class Notes (continued)

Your Notes

(Insert additional paper as needed.)

Practice:

☐   Complete the Vocabulary and Readiness Check on page 348.

☐   Next, complete any incomplete exercises below. Check and correct your work using the answers and references at the end of this section.

Review this example:	**Your turn:**

Review this example:

1. Write each percent as a fraction or mixed number in simplest form.

    a. 125%   b. $33\frac{1}{3}\%$

Replace the percent symbol with its equivalent,

$\frac{1}{100}$. Multiply, then simplify.

a. $125\% = 125 \cdot \frac{1}{100} = \frac{125}{100} = \frac{5 \cdot \overset{1}{\cancel{25}}}{4 \cdot \cancel{25}} = \frac{5}{4} \text{ or } \left(1\frac{1}{4}\right)$

b. $33\frac{1}{3}\% = 33\frac{1}{3} \cdot \frac{1}{100} = \frac{100}{3} \cdot \frac{1}{100} = \frac{\overset{1}{\cancel{100}} \cdot 1}{3 \cdot \underset{1}{\cancel{100}}} = \left(\frac{1}{3}\right)$

        Write as an improper fraction.

Your turn:

2. Write each percent as a fraction or mixed number in simplest form.

    a. 175%

    b. $10\frac{1}{3}\%$

Complete this example:	**Your turn:**

Complete this example:

3. Write the fraction $\frac{9}{20}$ as a percent.

When we multiply by 1, we are not changing the value of that number. Recall $100\% = 1$.

$\frac{9}{20} = \frac{9}{20} \cdot 100\% = \frac{9}{20} \cdot \frac{100}{1}\% = \frac{900}{20}\%$ Simplify.

$\frac{9}{20} = \left(\underline{\hspace{2cm}}\%\right)$

Your turn:

4. Write the fraction $\frac{7}{10}$ as a percent.

Section 5.3 Percents and Fractions

Review this example:

5. Write $\dfrac{1}{12}$ as a percent. Round to the nearest hundredth percent.

Multiply by 100%.

$$\dfrac{1}{12} = \dfrac{1}{12} \cdot 100\% = \dfrac{1}{12} \cdot \dfrac{100}{1}\% = \dfrac{100}{12}\%$$

Now, use long division to divide 100 by 12.

$$
\begin{array}{r}
8.333 \approx 8.33 \\
12\overline{)100.000} \\
-96 \\
\hline
40 \\
-36 \\
\hline
40 \\
-36 \\
\hline
40 \\
-36 \\
\hline
4
\end{array}
$$

Solution: $\boxed{\dfrac{1}{12} \approx 8.33\%}$

Remainder is 4. Bring 0 down.

Remainder is 4. Bring 0 down.

Remainder is 4. Notice the answer is a repeating decimal.

Your turn:

6. Write $\dfrac{4}{15}$ as a percent. Round to the nearest hundredth percent.

	Answer	Text Ref	Video Ref		Answer	Text Ref	Video Ref
1	a. $\dfrac{5}{4}$ or $1\dfrac{1}{4}$ b. $\dfrac{1}{3}$	Ex 3 & 4, p. 344		**4**	70%		Sec 5.3, 4/7
2	a. $\dfrac{7}{4}$ or $1\dfrac{3}{4}$ b. $\dfrac{31}{300}$		Sec 5.3, 2-3/7	**5**	$\approx 8.33\%$	Ex 9, p. 346	
3	45%	Ex 6, p. 345		**6**	$\approx 26.67\%$		Sec 5.3, 6/7

☐   **Next, insert your homework.** Make sure you attempt all exercises asked of you and show all work, as in the exercises above. Check your answers if possible. Clearly mark any exercises you were unable to correctly complete so that you may ask questions later. DO NOT ERASE YOUR INCORRECT WORK. THIS IS HOW WE UNDERSTAND AND EXPLAIN TO YOU YOUR ERRORS.

Section 5.4 Solving Percent Problems Using Equations

Before Class:

☐   Read the objectives on page 353.

☐   Read the **Helpful Hint** boxes on pages 353, 354, 355, 356 and 357.

☐   Complete the exercises:

      1. Fill in the table:

Percent	Decimal	Fraction
60%		
		$\dfrac{2}{5}$
	0.25	
12.5%		
		$\dfrac{5}{8}$
		$\dfrac{7}{50}$

During Class:

☐   **Write your class notes.** Neatly write down **all** examples shown as well as key terms or phrases with definitions. If not applicable or if you were absent, watch the Lecture Series (DVD) for this section and do the same (write down the examples shown as well as key terms or phrases). Insert more paper as needed.

Class Notes/Examples	**Your Notes**

Answers:

1) $60\% = 0.6 = \dfrac{3}{5}$; $40\% = 0.4 = \dfrac{2}{5}$; $25\% = 0.25 = \dfrac{1}{4}$; $12.5\% = 0.125 = \dfrac{1}{8}$; $62.5\% = 0.625 = \dfrac{5}{8}$; $14\% = 0.14 = \dfrac{7}{50}$

Section 5.4 Solving Percent Problems Using Equations

Class Notes (continued)	**Your Notes**

(Insert additional paper as needed.)

Section 5.4 Solving Percent Problems Using Equations

Practice:

☐   Complete the Vocabulary and Readiness Check on page 358.

☐   Next, complete any incomplete exercises below. Check and correct your work using the answers and references at the end of this section.

Review this example:
1. Translate to an equation. Do not solve.
 	
    38% of 200 is what number?
 	
38% of 200 is what number?
↓   ↓   ↓   ↓   ↓
38% · 200 = n
 	
The equation is: $38\% \cdot 200 = n$
	Your turn:
	2. Translate to an equation. Do not solve.
	
	    18% of 81 is what number?

Review this example:
3. Translate to an equation. Do not solve.
 	
    What percent of 85 is 34?
 	
What percent of 85 is 34?
↓   ↓ ↓ ↓
n · 85 = 34
 	
The equation is: $n \cdot 85 = 34$
	Your turn:
	4. Translate to an equation. Do not solve.
	
	    What percent of 80 is 3.8?

Complete this example:
5. 85% of 300 is what number?
 	
85% of 300 is what number?
↓   ↓   ↓   ↓   ↓
85% · 300 = n Translate to an equation.
0.85 · 300 = n Write 85% as 0.85.
       ? = n Multiply.
 	
  85% of 300 is _____.
	Your turn:
	6. 10% of 35 is what number?

Section 5.4 Solving Percent Problems Using Equations

Review this example:

7. 12% of what number is 0.6?

12% of what number is 0.6?
↓   ↓ ↓   ↓   ↓

| 12% | · | n | | = 0.6 | Translate and write |
| 0.12 | · | n | | = 0.6 | 12% as 0.12. |

$$\frac{0.12n}{0.12} = \frac{0.6}{0.12}$$ Divide both sides
by 0.12.

$$n = 5$$

12% of ⑤ is 0.6.

Your turn:

8. 1.2 is 12% of what number?

Complete this example:

9. 78 is what percent of 65?

78 is what percent of 65 ?
↓   ↓   ↓   ↓   ↓

78 = n · 65   Translate.

$$\frac{78}{?} = \frac{n \cdot 65}{?}$$ Divide both sides
by _____.

_____ = n or _____ % = n

⟨ 78 is _____ % of 65 . ⟩

Your turn:

10. 2.58 is what percent of 50?

	Answer	Text Ref	Video Ref		Answer	Text Ref	Video Ref
1	$38\% \cdot 200 = n$	Ex 4, p. 354		**6**	3.5		Sec 5.4, 5/6
2	$18\% \cdot 81 = n$		Sec 5.4, 1/6	**7**	5	Ex 9, p. 355	
3	$n \cdot 85 = 34$	Ex 6, p. 354		**8**	10		Sec 5.4, 4/6
4	$n \cdot 80 = 3.8$		Sec 5.4, 2/6	**9**	1.2 = n, 120% = n, 120%	Ex 12, p. 356	
5	255	Ex 8, p. 355		**10**	5.16%		Sec 5.4, 6/6

☐ **Next, insert your homework.** Make sure you attempt all exercises asked of you and show all work, as in the exercises above. Check your answers if possible. Clearly mark any exercises you were unable to correctly complete so that you may ask questions later. DO NOT ERASE YOUR INCORRECT WORK. THIS IS HOW WE UNDERSTAND AND EXPLAIN TO YOU YOUR ERRORS.

Section 5.5 Solving Percent Problems Using Proportions

Before Class:

☐   Read the objectives on page 361.

☐   Read the **Helpful Hint** boxes on pages 361, 363 and 365.

☐   Read the Percent Proportion Equation on page 361.

☐   Complete the exercises:

1. Fill in the chart.

Part of Proportion	How It's Identified
Percent	
Base	
Amount	

2. Fill in the blanks with **percent**, **base**, or **amount**.

In the statement "35% of what number is 84?", which part of the percent proportion is unknown? _____

During Class:

☐   **Write your class notes.**   Neatly write down **all** examples shown as well as key terms or phrases with definitions. If not applicable or if you were absent, watch the Lecture Series (DVD) for this section and do the same (write down the examples shown as well as key terms or phrases). Insert more paper as needed.

Class Notes/Examples	**Your Notes**

Answers: **1)** % or percent, Appears after *of*, Part compared to whole   **2)** base

Section 5.5 Solving Percent Problems Using Proportions

Class Notes (continued)	Your Notes

(Insert additional paper as needed.)

Section 5.5 Solving Percent Problems Using Proportions

Practice:

☐   Complete the Vocabulary and Readiness Check on page 366.

☐   Next, complete any incomplete exercises below. Check and correct your work using the
answers and references at the end of this section.

Review this example:
1. Translate to a proportion. Do not solve.

12% of what number is 47?

> *Recall:* base – appears after the word *of*
> amount – part compared to whole

12%   of   what   number   is   47?
↓   ↓   ↓
percent   base   amount

$$\frac{\text{amount}}{\text{base}} = \frac{\text{percent}}{100}$$ $\frac{47}{b} = \frac{12}{100}$

Your turn:
2. Translate to a proportion. Do not solve.

98% of 45 is what number?

Complete this example:
3. Translate to a proportion. Then solve.

20.8 is 40% of what number?

Identify the percent, base, and amount.

20.8   is   40%   of   what number?
↓   ↓   ↓
amount   percent   base

$\frac{20.8}{b} = \frac{40}{100}$ or $\frac{20.8}{b} = \frac{2}{5}$ Write the proportion.

Simplify.

$(20.8) \cdot 5 = b \cdot 2$ Set cross-products equal.

$104 = 2b$ Multiply.

$\frac{104}{2} = \frac{2b}{2}$ Divide both sides by 2.

$? = b$ Simplify.

20.8 is 40% of _____ .

Your turn:
4. Translate to a proportion. Then solve.

7.8 is 78% of what number?

Section 5.5 Solving Percent Problems Using Proportions

Complete this example:

5. Solve. What percent of 50 is 8?

| Your turn: |
| **6.** Solve. What percent of 6 is 2.7? |

Identify the percent, base, and amount.

What percent of 50 is 8?
↓   ↓   ↓
percent   base amount

$$\frac{\text{amount}}{\text{base}} = \frac{\text{percent}}{100}$$

$\dfrac{8}{50} = \dfrac{p}{100}$ or $\dfrac{4}{25} = \dfrac{p}{100}$ Write the proportion.

Simplify.

$\quad 4 \cdot 100 = 25 \cdot p$ Set cross-products equal.

$\quad 400 = 25p$ Multiply.

$\quad \dfrac{400}{?} = \dfrac{25p}{?}$ Divide both sides by

_____ .

_____ $= p$ Simplify.

So, $\left(\text{_____}\%\right)$ of 50 is 8 .

	Answer	Text Ref	Video Ref			Answer	Text Ref	Video Ref
1	$\dfrac{47}{b} = \dfrac{12}{100}$	Ex 1, p. 362			4	10		Sec 5.5, 3-4/6
2	$\dfrac{a}{45} = \dfrac{98}{100}$		Sec 5.5, 1/6		5	25; 16; 16%	Ex 10, p. 365	
3	52	Ex 9, p. 364			6	45%		Sec 5.5, 5/6

☐   **Next, insert your homework.** Make sure you attempt all exercises asked of you and show all work, as in the exercises above. Check your answers if possible. Clearly mark any exercises you were unable to correctly complete so that you may ask questions later. DO NOT ERASE YOUR INCORRECT WORK. THIS IS HOW WE UNDERSTAND AND EXPLAIN TO YOU YOUR ERRORS.

Section 5.6 Applications of Percent

Before Class:

☐   Read the objectives on page 371.

☐   Read the **Helpful Hint** box on page 375.

☐   Read the Percent of Increase box on page 374 and the Percent of Decrease box on page 375.

☐   Complete the exercises:

1.   Write 4% as a decimal. _____

2.   Fill in the blanks with **percent**, **base**, or **amount**.

In the statement "864 is 32% of 2700," the number 864 is called the _____ ,

32 is called the _____ , and 2700 is the _____.

3.   A group of 1500 employees is reduced to 1230 employees. What is the amount of decrease?

4.   To determine the percent of increase or percent of decrease, we divide by the

_____ .

During Class:

☐   **Write your class notes.** Neatly write down **all** examples shown as well as key terms or phrases with definitions. If not applicable or if you were absent, watch the Lecture Series (DVD) for this section and do the same (write down the examples shown as well as key terms or phrases). Insert more paper as needed.

Class Notes/Examples	**Your Notes**

Answers: **1)** 0.04   **2)** amount, percent, base   **3)** 270   **4)** original amount

Section 5.6 Applications of Percent

Class Notes (continued)	**Your Notes**

(Insert additional paper as needed.)

Practice:

☐   Complete any incomplete exercises below. Check and correct your work using the answers and references at the end of this section.

Review this example:

1. Mr. Buccaran, the principal at Slidell High School, counted 31 freshmen absent during a particular day. If this is 4% of the total number of freshmen, how many freshmen are there at Slidell High School?

31 is the amount, 4 is the percent, and n is the base.

| *Method 1: Translate to an equation.* |

31 is 4% of what number?

↓   ↓ ↓   ↓   ↓

31 = 4% · n

$31 = 0.04n$ Write 4% as a decimal.

$$\frac{31}{0.04} = \frac{0.04n}{0.04}$$ Divide both sides by 0.04.

$775 = n$ Simplify.

There are 775 freshmen at Slidell High School.

| *Method 2: Translate to a proportion.* |

31 is 4% of what number?

amount → $\dfrac{31}{b} = \dfrac{4}{100}$ ← percent
base →

$$\frac{31}{b} = \frac{4}{100}$$

$3100 = 4b$ Cross products are equal.

$$\frac{3100}{4} = \frac{4b}{4}$$ Divide both sides by 4.

$775 = b$ Simplify.

Your turn:

2. A family paid $26,250 as a down payment for a home. If this represents 15% of the price of the home, find the price of the home. Use both methods.

Section 5.6 Applications of Percent

Review this example:

3. In response to a decrease in sales, a company with 1500 employees reduces the number of employees to 1230. What is the percent decrease?

Find the amount of decrease. $1500 - 1230 = 270$

$$\frac{\text{amount of decrease}}{\text{original amount}} = \frac{270}{1500} = 0.18 = 18\%$$

The number of employees decreased by 18%.

Your turn:

4. There are 150 calories in a cup of whole milk and only 84 calories in a cup of skim milk. In switching to skim milk, find the percent decrease in the number of calories per cup.

Review this example:

5. The number of applications for a mathematics scholarship at Yale increased from 34 to 45 in one year. What is the percent increase? Round to the nearest whole percent.

Find the amount of increase. $45 - 34 = 11$

$$\frac{\text{amount of increase}}{\text{original amount}} = \frac{11}{34} \approx 0.32 \approx 32\%$$

The number of applications increased by about 32%.

Your turn:

6. In 1940, the average size of a privately owned farm in the United States was 174 acres. In a recent year, the average size of a privately owned farm in the United States had increased to 449 acres. What is this percent increase? (*Source*: National Agricultural Statistics Service)

	Answer	Text Ref	Video Ref		Answer	Text Ref	Video Ref
1	775 freshmen	Ex 3, p. 372-3		4	44%		Sec 5.6, 2/3
2	$175,000		Sec 5.6, 1/3	5	≈ 32%	Ex 5, p. 375	
3	18%	Ex 6, p. 375		6	158.0%		Sec 5.6, 3/3

☐   **Next, insert your homework.** Make sure you attempt all exercises asked of you and show all work, as in the exercises above. Check your answers if possible. Clearly mark any exercises you were unable to correctly complete so that you may ask questions later. DO NOT ERASE YOUR INCORRECT WORK. THIS IS HOW WE UNDERSTAND AND EXPLAIN TO YOU YOUR ERRORS.

Section 5.7 Percent and Problem Solving: Sales Tax, Commission, and Discount

Before Class:

☐ Read the objectives on page 382.

☐ Read the Sales Tax and Total Price box on page 382 and the Commission box on page 383.

☐ Read the Discount and Sale Price box on page 384.

☐ Complete the exercises:

1. If a number is increased by 100%, how does the increased number compare with the original number? Explain your answer.

2. In your own words, explain what is wrong with the following statement.
    "Last year we had 80 students attend. This year we have a 50% increase or a total of 160 students attend."

During Class:

☐ **Write your class notes.** Neatly write down **all** examples shown as well as key terms or phrases with definitions. If not applicable or if you were absent, watch the Lecture Series (DVD) for this section and do the same (write down the examples shown as well as key terms or phrases). Insert more paper as needed.

Class Notes/Examples	**Your Notes**

Answers: **1)** The increased number is double the original number.   **2)** If there is a 50% increase, then $.50(80) = 40$ more students than last year. $80 + 40 = 120$; If there was a 100% increase, the 160 students given would have been correct.

Section 5.7 Percent and Problem Solving: Sales Tax, Commission, and Discount

Class Notes (continued)	Your Notes

(Insert additional paper as needed.)

Section 5.7 Percent and Problem Solving: Sales Tax, Commission, and Discount

Practice:

☐   Complete the Vocabulary and Readiness Check on page 386.

☐   Next, complete any incomplete exercises below. Check and correct your work using the answers and references at the end of this section.

Review this example:

1. Find the sales tax and the total price on the purchase of an $85.50 atlas in a city where the sales tax rate is 7.5%.

sales tax = tax rate · purchase price
↓   ↓   ↓

| sales tax | =   7.5%   · $85.50

= 0.075·$85.50   Write 7.5% as a decimal. Multiply and round to nearest cent.
≈ $6.41

total price = purchase price + sales tax
↓   ↓   ↓

| total price | = $85.50   + $6.41
= $91.91

The sales tax on $85.50 is ($6.41) and the total price is ($91.91.)

Your turn:

2. The sales tax on the purchase of a futon is $24.25. If the tax rate is 5%, find the purchase price of the futon.

Review this example:

3. A salesperson earned $1560 for selling $13,000 worth of electronics equipment. Find the commission rate.

commission = commission rate · sales
↓   ↓   ↓
$1560   =   r   · $13,000

$\dfrac{1560}{13,000} = r$ Divide both sides by 13,000.

0.12 = r   Simplify.
12% = r   Write 0.12 as a percent.

The commission rate was (12%) of the sale.

Your turn:

4. A salesperson earned a commission of $1380.40 for selling $9860 worth of paper products. Find the commission rate.

135

Section 5.7 Percent and Problem Solving: Sales Tax, Commission, and Discount

Complete this example:

5. An electric rice cooker that normally sells for $65 is on sale for 25% off. What is the amount of discount and what is the sale price?

amount of discount = discount rate · original price
$$\downarrow \qquad\qquad \downarrow \qquad\qquad \downarrow$$

| amount of discount | = | 25% | · | $65 |

$$= \ 0.25 \cdot \$65 \qquad \text{Write 25\%}$$
$$= \ \$16.25 \qquad\quad \text{as a decimal}$$
$$\qquad\qquad\qquad\qquad \text{and multiply.}$$

sale price = original price − discount
$$\downarrow \qquad\qquad \downarrow \qquad\qquad \downarrow$$

| sale price | = | $65 | − | $16.25 |

The sale price is _____.

Your turn:

6. A $300 fax machine is on sale for 15% off. Find the amount of discount and the sale price.

	Answer	Text Ref	Video Ref		Answer	Text Ref	Video Ref
1	sales tax: $6.41, total: $91.91	Ex 1, p. 382		4	14%		Sec 5.7, 2/3
2	$485		Sec 5.7, 1/3	5	$48.75	Ex 5, p. 385	
3	12%	Ex 4, p. 384		6	$45; $255		Sec 5.7, 3/3

☐ **Next, insert your homework.** Make sure you attempt all exercises asked of you and show all work, as in the exercises above. Check your answers if possible. Clearly mark any exercises you were unable to correctly complete so that you may ask questions later. DO NOT ERASE YOUR INCORRECT WORK. THIS IS HOW WE UNDERSTAND AND EXPLAIN TO YOU YOUR ERRORS.

Section 5.8 Percent and Problem Solving: Interest

Before Class:

☐   Read the objectives on page 389.

☐   Read the **Helpful Hint** box on page 391.

☐   Read the Simple Interest box on page 389 and the Compound Interest Formula on page 391.

☐   Complete the exercises:

1. Write 5.3% as a decimal. _____

2. a. What part of a year is 6 months? _____

    b. What part of a year is 8 months? _____

3. In the simple interest formula the rate is understood to be per _____ .

4. Simplify $1800(1.02)^3$ using a calculator. Round your answer to two decimal places.

During Class:

☐   **Write your class notes.** Neatly write down **all** examples shown as well as key terms or phrases with definitions. If not applicable or if you were absent, watch the Lecture Series (DVD) for this section and do the same (write down the examples shown as well as key terms or phrases). Insert more paper as needed.

Class Notes/Examples	Your Notes

Answers: **1)** 0.053   **2)** a. $\frac{1}{2}$ year b. $\frac{2}{3}$ year   **3)** year   **4)** ≈ 1910.17

Section 5.8 Percent and Problem Solving: Interest

Class Notes (continued)	Your Notes

(Insert additional paper as needed.)

Section 5.8 Percent and Problem Solving: Interest

Practice:

☐   Complete the Vocabulary and Readiness Check on page 394.

☐   Next, complete any incomplete exercises below. Check and correct your work using the answers and references at the end of this section.

Review this example:

1. a. Find the simple interest after 2 years on $500 at an interest rate of 12% .

a. $P = \$500$, $R = 12\%$, and $T = 2$ years

Simple Interest = Principal · Rate · Time
$I = P \cdot R \cdot T$

$I = \$500 \cdot 12\% \cdot 2$

$I = \$500 \cdot (0.12) \cdot 2$ Write 12% as a decimal.

$I = \$120$ Multiply.

The simple interest is $120.

   b. Ivan Borski borrowed $2400 at 10% simple interest for 8 months to buy a used Toyota Corolla. Find the simple interest he paid.

b. $P = \$2400$, $R = 10\%$, and $T = \dfrac{8}{12} = \dfrac{2}{3}$ year

$I = \$2400 \cdot 10\% \cdot \dfrac{2}{3}$

$I = \$2400 \cdot (0.10) \cdot \dfrac{2}{3}$ Write 10% as a decimal.

$I = \$160$ Multiply.

The simple interest is $160.

Your turn:

2. A company borrows $162,500 for 5 years at a simple interest rate of 12.5%.

   a. Find the interest paid on the loan.

   b. Find the total amount paid back.

Section 5.8 Percent and Problem Solving: Interest

Review this example:

3. $4000 is invested at 5.3% compounded quarterly for 10 years. Find the total amount at the end of 10 years.

> **Use the compound interest formula:**
>
> $$A = P\left(1 + \frac{r}{n}\right)^{n \cdot t}$$

Compounded quarterly means 4 times a year, so $n = 4$. $P = \$4000$, $r = 5.3\%$, and $t = 10$ years.

$$A = P\left(1 + \frac{r}{n}\right)^{n \cdot t}$$

$$A = 4000\left(1 + \frac{0.053}{4}\right)^{4 \cdot 10}$$

$$A = 4000(1.01325)^{40}$$ Evaluate $(1.01325)^{40}$ first. Then multiply by 4000.

$$A \approx 6772.12$$

The total amount after 10 years is about $6772.12 .

Your turn:

4. Find the total amount in the compound interest account if $6150 is compounded semiannually at a rate of 14% for 15 years.

	Answer	Text Ref	Video Ref		Answer	Text Ref	Video Ref
1	a. $120 b. $160	Ex 1, p. 389 Ex 2, p. 389		**3**	$\approx \$6772.12$	Ex 5, p. 392	
2	a. $101,562.50 b. $264,062.50		Sec 5.8, 1/3	**4**	$\approx \$46,815.37$		Sec 5.8, 2/3

☐   **Next, insert your homework.** Make sure you attempt all exercises asked of you and show all work, as in the exercises above. Check your answers if possible. Clearly mark any exercises you were unable to correctly complete so that you may ask questions later. DO NOT ERASE YOUR INCORRECT WORK. THIS IS HOW WE UNDERSTAND AND EXPLAIN TO YOU YOUR ERRORS.

Preparing for the Chapter 5 Test

Start preparing for your Chapter 5 Test as soon as possible. Pay careful attention to any instructor discussion about this test, especially discussion on what sections you will be responsible for, etc.

☐ Work the Chapter 5 Vocabulary Check on page 397.

☐ Read both columns (Definitions and Concepts, and Examples) of the Chapter 5 Highlights starting on page 398.

☐ Read your Class Notes/Examples for each section covered on your Chapter 5 Test. Look for any unresolved questions you may have.

☐ Complete as many of the Chapter 5 Review exercises as possible (starting on page 402). Remember, the odd answers are in the back of your text.

☐ **Most important:** Place yourself in "test" conditions (see below) and work the Chapter 5 Test (pages 406 – 407) as a practice test the day before your actual test. To honestly assess how you are doing, try the following:
   - Work on a few blank sheets of paper.
   - Give yourself the same amount of time you will be given for your actual test.
   - Complete this Chapter 5 Practice Test without using your notes or your text.
   - If you have any time left after completing this practice test, check your work and try to find any errors on your own.
   - Once done, use the back of your book to check ALL answers.
   - Try to correct any errors on your own.
   - Use the Chapter Test Prep Video (CTPV) to correct any errors you were unable to correct on your own. You can find these videos in the Interactive DVD Lecture Series, in MyMathLab, and on YouTube. Search MartinGayDevMath and click "Channels."

I wish you the best of luck….Elayn Martin-Gay

Before Class:

☐   Read the objectives on page 411.

☐   Read the **Helpful Hint** box on page 412.

☐   Read the paragraph above the Parallel Lines Cut by a Transversal box on page 416.

☐   Complete the exercises:

    1.   The measure of $\angle A$ added to the measure of $\angle B$ totals 90°.
       If the measure of $\angle A = 40°$, find the measure of $\angle B$. _____

    2.   The measure of $\angle S$ added to the measure of $\angle T$ totals 180°.
       If the measure of $\angle S = 40°$, find the measure of $\angle T$. _____

    3.   The measure of $\angle U$ added to the measure of $\angle V$ totals 180°.
       If the measure of $\angle V = 55°$, find the measure of $\angle U$. _____

During Class:

☐   **Write your class notes.** Neatly write down **all** examples shown as well as key terms or phrases with definitions. If not applicable or if you were absent, watch the Lecture Series (DVD) for this section and do the same (write down the examples shown as well as key terms or phrases). Insert more paper as needed.

Class Notes/Examples	Your Notes

Answers: **1)** $m\angle B = 50°$ **2)** $m\angle T = 140°$ **3)** $\angle U = 125°$

Section 6.1 Lines and Angles

Class Notes (continued)	Your Notes

(Insert additional paper as needed.)

Practice:

☐   Complete the Vocabulary and Readiness Check on page 417.

☐   Next, complete any incomplete exercises below. Check and correct your work using the answers and references at the end of this section.

Review this example:	**Your turn:**
1. Identify each figure as a line, a ray, a line segment, or an angle. Then name the figure using the given points.	**2.** Identify each figure as a line, a ray, a line segment, or an angle. Then name the figure using the given points.

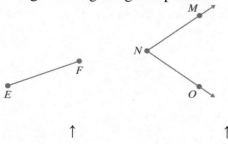

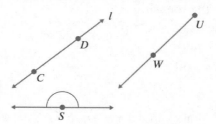

The figure has two endpoints.

The figure has two rays with a common endpoint.

It is line segment
EF or $\overline{EF}$.

The angle is
$\angle MNO$, $\angle ONM$,
or $\angle N$.

Review this example:	**Your turn:**
3. Classify each angle as acute, right, obtuse, or straight.	**4.** Classify each angle as acute, right, obtuse, or straight.

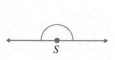

$\angle S$ is a (straight) angle. It measures 180°.

$\angle T$ is an (acute) angle. It measures between 0° and 90°.

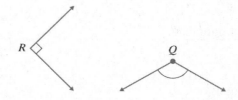

Section 6.1 Lines and Angles

Complete this example:

5. a. Find the complement of a 48° angle.

b. Find the supplement of a 107° angle.

a. Complementary: Two angles whose sum is 90°

90° − 48° = _____°

b. Supplementary: Two angles whose sum is 180°

180° − 107° = _____°

Your turn:

6. a. Find the complement of a 23° angle.

b. Find the supplement of a 17° angle.

Review this example:

7. Find the measures of angles *x*, *y*, and *z* if *m* ∥ *n* and ∠*w* = 100°.

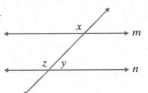

∠*x* and ∠*w* are vertical angles, so m∠*x* = 100°.

∠*x* and ∠*z* are corresponding angles, so m∠*z* = 100°.

∠*z* and ∠*y* are supplementary angles, so m∠*y* = 180° − 100° = 80°.

Your turn:

8. Find the measures of angles *x*, *y*, and *z* if *m* ∥ *n*.

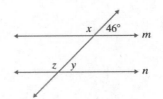

	Answer	Text Ref	Video Ref		Answer	Text Ref	Video Ref
1	line segment EF or $\overline{EF}$; angle, ∠*MNO*, ∠*ONM*, or ∠*N*	Ex 1b,c p. 412		5	a. 42° b. 73°	Ex 4, p. 414 Ex 5, p. 415	
2	line, line CD or line *l* or $\overleftrightarrow{CD}$; ray, ray *UW* or $\overrightarrow{UW}$		Sec 6.1, 1 – 2/9	6	a. 67° b. 163°		Sec 6.1, 5-6/9
3	straight; acute	Ex 3b,c p. 413		7	m∠*x* = 100° m∠*y* = 80° m∠*z* = 100°	Ex 8, p. 416	
4	right; obtuse		Sec 6.1, 3-4/9	8	m∠*x* = 134° m∠*y* = 46° m∠*z* = 134°		Sec 6.1, 9/9

☐ **Next, insert your homework.** Make sure you attempt all exercises asked of you and show all work, as in the exercises above. Check your answers if possible. Clearly mark any exercises you were unable to correctly complete so that you may ask questions later. DO NOT ERASE YOUR INCORRECT WORK. THIS IS HOW WE UNDERSTAND AND EXPLAIN TO YOU YOUR ERRORS.

Section 6.2 Plane Figures and Solids

Before Class:

☐ Read the objectives on page 421.

☐ Look over the Polygon Chart on page 421, the examples of triangles on page 422, the examples of quadrilaterals on page 423 and the examples of solid figures on page 424-425.

☐ Complete these exercises:

1. The pentagon is a polygon with _____ sides.

2. A square has _____ sides, a triangle has _____ sides, and a trapezoid has _____ sides.

3. Name two characteristics of a scalene triangle.

4. The sum of the measures of the angles of a triangle is _____.

During Class:

☐ Write your class notes. Neatly write down all examples shown as well as key terms or phrases with definitions. If not applicable or if you were absent, watch the Lecture Series (DVD) for this section and do the same (write down the examples shown as well as key terms or phrases). Insert more paper as needed.

Class Notes/Examples	**Your Notes**

Answers: **1)** five **2)** four, three, four **3)** no sides are the same length; no angles have the same measure **4)** 180°

Section 6.2 Plane Figures and Solids

Class Notes (continued)	**Your Notes**

(Insert additional paper as needed.)

Section 6.2 Plane Figures and Solids

Practice:

☐   Next, complete any incomplete exercises below. Check and correct your work using the answers and references at the end of this section.

Review this example:
1. Find the measure of $\angle a$.

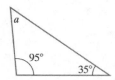

The sum of the measures of the angles of a triangle is 180°.

$m\angle a = 180° - 95° - 35° = \boxed{50°}$

Your turn:
2. Find the measure of $\angle x$.

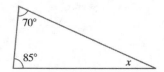

Review this example:
3. Find the measure of $\angle b$.

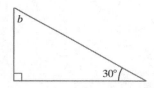

The measure of the right angle is 90°.

$m\angle b = 180° - 90° - 30° = \boxed{60°}$

Your turn:
4. Find the measure of $\angle x$.

Review this example:
5. Find the unknown diameter or radius in each figure.

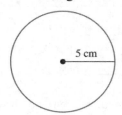

The diameter is twice the radius.

$d = 2 \cdot r = 2 \cdot 5 = \boxed{10\,cm}$

Your turn:
6. Find the unknown diameter or radius in each figure.

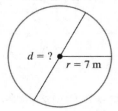

149

Section 6.2 Plane Figures and Solids

Complete this example:

7. Identify each solid.

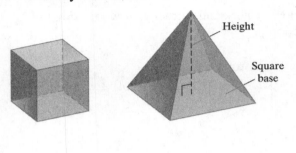

Height

Square base

a._____

b._____

Your turn:

8. Identify each solid.

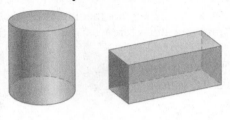

a._____

b._____

Review this example:

9. Find the radius of the sphere.

36 ft

The radius is half the diameter.

$$r = \frac{d}{2} = \frac{36}{2} = 18 \; ft$$

Your turn:

10. Find the radius of the sphere.

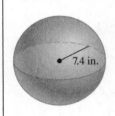

7.4 in.

	Answer	Text Ref	Video Ref		Answer	Text Ref	Video Ref
1	$m\angle a = 50°$	Ex 1, p. 422		6	14 m		Sec 6.2, 5/7
2	$m\angle x = 25°$		Sec 6.2, 1/7	7	a. cube b. pyramid	Chart p.424	
3	$m\angle b = 60°$	Ex 2, p. 422		8	a. cylinder b. rectangular solid		Sec 6.2, 6-7/7
4	$m\angle x = 40°$		Sec 6.2, 2/7	9	18 ft	Ex 4, p.425	
5	10 cm	Ex 3, p.424		10	14.8 in.		p.428, #51 no video

☐   **Next, insert your homework.** Make sure you attempt all exercises asked of you and show all work, as in the exercises above. Check your answers if possible. Clearly mark any exercises you were unable to correctly complete so that you may ask questions later. DO NOT ERASE YOUR INCORRECT WORK. THIS IS HOW WE UNDERSTAND AND EXPLAIN TO YOU YOUR ERRORS.

Section 6.3 Perimeter

Before Class:

☐ Read the objectives on page 430.

☐ Read the Perimeter boxes of the following special figures: rectangle page 430, square page 431, and triangle page 431; and the Circumference of a Circle box on page 433.

☐ Complete the exercises:

1. The perimeter of a polygon is the _____ of the lengths of its sides.

2. The distance around a circle is called the _____ .

3. If the diameter of a circle is 10 ft, what is the radius of the circle? _____

4. $\pi \approx$ _____ (two decimal places) and $\pi \approx$ _____ (in fraction form).

During Class:

☐ **Write your class notes.** Neatly write down all examples shown as well as key terms or phrases with definitions. If not applicable or if you were absent, watch the Lecture Series (DVD) for this section and do the same (write down the examples shown as well as key terms or phrases). Insert more paper as needed.

Class Notes/Examples	**Your Notes**

Answers: **1)** sum   **2)** circumference   **3)** 5 ft   **4)** $3.14, \dfrac{22}{7}$

Section 6.3 Perimeter

Class Notes (continued)	Your Notes

(Insert additional paper as needed.)

Practice:

☐   Complete the Vocabulary and Readiness Check on page 435.

☐   Next, complete any incomplete exercises below. Check and correct your work using the
        answers and references at the end of this section.

Review this example:	**Your turn:**
1. Find the perimeter of the rectangle.	**2.** Find the perimeter of the figure.

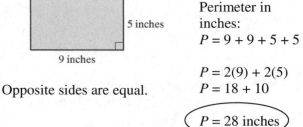

Perimeter in inches:

$P = 9 + 9 + 5 + 5$

$P = 2(9) + 2(5)$

$P = 18 + 10$

Opposite sides are equal.

$\boxed{P = 28 \text{ inches}}$

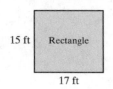

Review this example:

3. Find the perimeter of a triangle if the sides are 3 inches, 7 inches, and 6 inches.

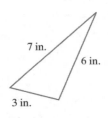

To find the perimeter, find the sum of the three sides of the triangle.

$P = 7 \text{ in.} + 6 \text{ in.} + 3 \text{ in.}$

$\boxed{P = 16 \text{ in.}}$

Your turn:

4. Find the perimeter of the figure.

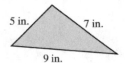

Review this example:

5. Find the perimeter of the trapezoid.

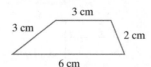

To find the perimeter, find the sum of the lengths of the four sides.

Perimeter in cm:

$P = 3 \text{ cm} + 2 \text{ cm} + 6 \text{ cm} + 3 \text{ cm} = \boxed{14 \text{ cm}}$

Your turn:

6. Find the perimeter of the figure.

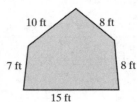

153

Section 6.3 Perimeter

Review this example:

7. Find the perimeter of the room shown.

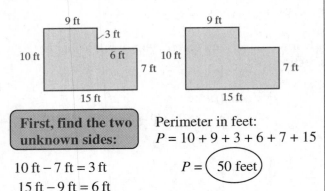

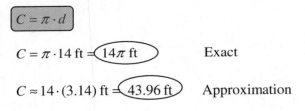

First, find the two unknown sides:

$10 \text{ ft} - 7 \text{ ft} = 3 \text{ ft}$

$15 \text{ ft} - 9 \text{ ft} = 6 \text{ ft}$

Perimeter in feet:

$P = 10 + 9 + 3 + 6 + 7 + 15$

$P = \boxed{50 \text{ feet}}$

Your turn:

8. Find the perimeter of the figure.

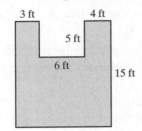

Review this example:

9. Mary Catherine Dooley plans to install a border of new tiling around the circumference of her circular spa. If her spa has a diameter of 14 feet, find its exact circumference. Then, use 3.14 for π to approximate the circumference.

$C = \pi \cdot d$

$C = \pi \cdot 14 \text{ ft} = \boxed{14\pi \text{ ft}}$ Exact

$C \approx 14 \cdot (3.14) \text{ ft} = \boxed{43.96 \text{ ft}}$ Approximation

Your turn:

10. Find the circumference of the circle. Give the exact circumference and then an approximation using π.

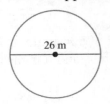

	Answer	Text Ref	Video Ref		Answer	Text Ref	Video Ref
1	28 in.	Ex 1, p. 430		6	48 ft		Sec 6.3, 3/6
2	64 ft		Sec 6.3, 1/6	7	50 ft	Ex 6, p.432	
3	16 in.	Ex 4, p. 432		8	66 ft		Sec 6.3, 5/6
4	21 in.		Sec 6.3, 2/6	9	14π ft, 43.96 ft	Ex 8, p.434	
5	14 cm	Ex 5, p. 432		10	26π m , 81.64 m		Sec 6.3, 6/6

☐   **Next, insert your homework.** Make sure you attempt all exercises asked of you and show all work, as in the exercises above. Check your answers if possible. Clearly mark any exercises you were unable to correctly complete so that you may ask questions later. DO NOT ERASE YOUR INCORRECT WORK. THIS IS HOW WE UNDERSTAND AND EXPLAIN TO YOU YOUR ERRORS.

Section 6.4 Area

Before Class:

☐ Read the objectives on page 440.

☐ Read the **Helpful Hint** boxes on pages 441 and 442.

☐ Look over the Area Formulas of Common Geometric Figures on page 440 and the Area Formula of a Circle on page 443.

☐ Complete the exercises:

1. Which figure has the greater distance around? (Use 3.14 for π.)

   a. A square with side length 3 inches   b. A circle with diameter 4 inches

2. Which figure has the greater distance around? (Use 3.14 for π)

   a. A circle with diameter 7 inches   b. A square with side length 7 inches

During Class:

☐ **Write your class notes.** Neatly write down **all** examples shown as well as key terms or phrases with definitions. If not applicable or if you were absent, watch the Lecture Series (DVD) for this section and do the same (write down the examples shown as well as key terms or phrases). Insert more paper as needed.

Class Notes/Examples	Your Notes

Answers: **1**) a. 12 in.   b. 12.56 in. (greater one)   **2**) a. 21.98 in.   b. 28 in. (greater one)

Section 6.4 Area

Class Notes (continued)	**Your Notes**

(Insert additional paper as needed.)

Practice:

☐   Next, complete any incomplete exercises below. Check and correct your work using the answers and references at the end of this section.

Review this example:
1. Find the area of the parallelogram.

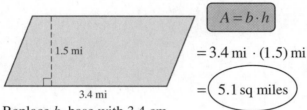

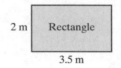

$A = b \cdot h$

$= 3.4 \text{ mi} \cdot (1.5) \text{ mi}$

$= \boxed{5.1 \text{ sq miles}}$

Replace b, base with 3.4 cm.
Replace h, height with 1.5

Your turn:
2. Find the area of the figure.

2 m   Rectangle

3.5 m

Review this example:
3. Find the area of the triangle.

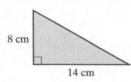

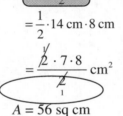

$A = \frac{1}{2} bh$

$= \frac{1}{2} \cdot 14 \text{ cm} \cdot 8 \text{ cm}$

$= \frac{\cancel{2} \cdot 7 \cdot 8}{\cancel{2}} \text{ cm}^2$

$A = 56$ sq cm

Replace b, base with 14 cm.
Replace h, height with 8 cm.

Your turn:
4. Find the area of the figure.

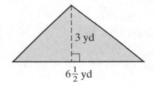

3 yd

$6\frac{1}{2}$ yd

Complete this example:
5. Find the area of the figure.

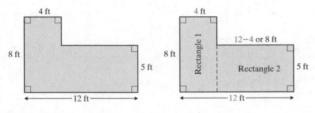

Rectangle 1: $A = l\,w = 8$ feet $\cdot$ 4 feet $= 32$ sq ft
Rectangle 2: $A = l\,w = 8$ feet $\cdot$ 5 feet $= 40$ sq ft

Total Area = Area Rectangle 1 + Area Rectangle 2

$= 32$ sq ft $+ 40$ sq ft $= \boxed{\underline{\hspace{1.5cm}}}$ sq ft

Your turn:
6. Find the area of the figure.

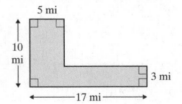

5 mi

10 mi

17 mi

3 mi

157

Section 6.4 Area

Complete this example:

7. Find the exact area of the figure. Then approximate the area using 3.14 for π.

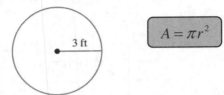

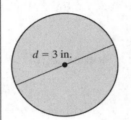

$$A = \pi r^2$$

Exact:

$A = \pi r^2 = \pi \cdot (3 \text{ feet})^2$

$A = \boxed{9\pi \text{ square feet}}$

Approximate:

$A \approx 9 \cdot (3.14)$ square feet

$A \approx \underline{\hspace{2cm}}$ square feet

Your turn:

8. Find the exact area of the circle. Then approximate using 3.14 for π.

	Answer	Text Ref	Video Ref		Answer	Text Ref	Video Ref
1	5.1 sq mi	Ex 2, p. 441		5	72 sq ft	Ex 3, p. 441 - 442	
2	7 sq m		Sec 6.4, 1/5	6	86 sq mi		Sec 6.4, 3/5
3	56 sq cm	Ex 1, p. 441		7	9π sq ft ≈ 28.26 sq ft	Ex 4, p. 443	
4	$9\frac{3}{4}$ sq yd		Sec 6.4, 2/5	8	2.25π sq in. ≈ 7.065 sq in.		Sec 6.4, 4/5

☐ **Next, insert your homework.** Make sure you attempt all exercises asked of you and show all work, as in the exercises above. Check your answers if possible. Clearly mark any exercises you were unable to correctly complete so that you may ask questions later. DO NOT ERASE YOUR INCORRECT WORK. THIS IS HOW WE UNDERSTAND AND EXPLAIN TO YOU YOUR ERRORS.

Before Class:

☐ Read the objectives on page 450.

☐ Read the **Helpful Hint** box on page 451.

☐ Look over the Volume Formulas of Common Solids on page 450 – 451.

☐ Complete the exercises:

Given the following situations, tell whether you are more likely to be concerned with area or perimeter.

1. ordering fencing to fence a yard _____

2. ordering grass seed to plant in a yard _____

3. buying carpet to install in a room _____

4. buying gutters to install on a house _____

5. ordering baseboards to install in a room _____

During Class:

☐ **Write your class notes.** Neatly write down **all** examples shown as well as key terms or phrases with definitions. If not applicable or if you were absent, watch the Lecture Series (DVD) for this section and do the same (write down the examples shown as well as key terms or phrases.) Insert more paper as needed.

Class Notes/Examples	**Your Notes**

Answers: **1)** perimeter   **2)** area   **3)** area   **4)** perimeter **5)** perimeter

Section 6.5 Volume

Class Notes (continued)	Your Notes

(Insert additional paper as needed.)

Practice:

☐   Complete the Vocabulary and Readiness Check on page 454.

☐   Next, complete any incomplete exercises below. Check and correct your work using the answers and references at the end of this section.

Review this example:

1. Find the volume of the figure.

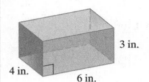

$$V = l \cdot w \cdot h$$

Let $h = 3$ in., $l = 12$ in., and $w = 6$ in.

$= 12$ inches $\cdot$ 6 inches $\cdot$ 3 inches $=$ 216 cu in.

The volume of the figure is 216 cubic inches.

Helpful Hint

Volume is always measured in cubic units.

Your turn:

2. Find the volume of the figure.

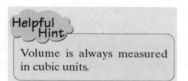

3 in.

4 in.

6 in.

Review this example:

3. Find the volume of the figure.
    Give an exact and an approximate answer.

Use $\dfrac{22}{7}$ for π.

3 in.

Exact:

$$V = \frac{4}{3} \cdot \pi \cdot r^3$$

$$= \frac{4}{3} \cdot \pi (3)^3$$

$$= \frac{4}{3} \cdot \pi (27)$$

$$= \frac{4 \cdot \pi \cdot \cancel{3}^{1} \cdot 9}{\cancel{3}_{1}}$$

$$= 36\pi \; cu.in.$$

Approximate:

$$V = 36\pi \; \text{cu.in.}$$

$$\approx 36 \cdot \frac{22}{7} \text{cu.in.}$$

$$\approx \frac{36 \cdot 22}{7} \text{cu.in.}$$

$$\approx \frac{792}{7} \text{cu.in.}$$

$$\approx \; 113\frac{1}{7}\text{cu.in.}$$

Your turn:

4. Find the volume of the figure. Use $\dfrac{22}{7}$ for π.

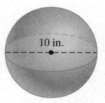

10 in.

Section 6.5 Volume

Review this example:

5. Find the volume of the figure. $V = \pi r^2 h$

Use $\dfrac{22}{7}$ for π.

$$V \approx \frac{22}{7} \cdot \left(\frac{7}{2}\text{in.}\right)^2 \cdot 6\text{in.}$$

$$\approx \frac{22}{7} \cdot \frac{49}{4}\,\text{sq.in.} \cdot 6\text{in.}$$

$$\approx \frac{\overset{1}{\cancel{2}} \cdot 11 \cdot \overset{1}{\cancel{7}} \cdot 7 \cdot \overset{1}{\cancel{2}} \cdot 3}{\underset{1}{\cancel{7}} \cdot \underset{1}{\cancel{2}} \cdot \underset{1}{\cancel{2}}}\,\text{cu.in.}$$

$$\approx \boxed{231\,\text{cu.in.}}$$

$3\tfrac{1}{2}$ in.

6 in.

Your turn:

6. Find the volume of the figure.

Use $\dfrac{22}{7}$ for π.

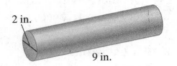
2 in.

9 in.

Review this example:

7. Approximate the volume of the cone:

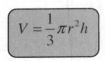
$V = \dfrac{1}{3}\pi r^2 h$

$$V = \frac{1}{3}\pi \cdot r^2 \cdot h$$

$$\approx \frac{1}{3} \cdot 3.14 \cdot (3\text{cm})^2 \cdot 14\text{cm}$$

$$\approx \boxed{131.88\,\text{cu.in.}}$$

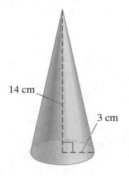
14 cm

3 cm

Your turn:

8. Find the exact volume of a waffle ice cream cone with a 3-in. diameter and a height of 7 inches.

	Answer	Text Ref	Video Ref		Answer	Text Ref	Video Ref
1	216 cu in.	Ex 1, p. 451		**5**	231 cu.in.	Ex 3, p.452	
2	72 cu in.		Sec 6.5, 1/5	**6**	$28\tfrac{2}{7}$ cu in.		Sec 6.5, 3/5
3	36π cu.in. $113\tfrac{1}{7}$ cu in.	Ex 2, p. 451		**7**	≈ 131.88 cu.in.	Ex 4, p.453	
4	$523\tfrac{17}{21}$ cu in.		Sec 6.5, 2/5	**8**	5.25π cu.in.		Sec 6.5, 5/5

☐   **Next, insert your homework.** Make sure you attempt all exercises asked of you and show all work, as in the exercises above. Check your answers if possible. Clearly mark any exercises you were unable to correctly complete so that you may ask questions later. DO NOT ERASE YOUR INCORRECT WORK. THIS IS HOW WE UNDERSTAND AND EXPLAIN TO YOU YOUR ERRORS.

Section 6.6 Square Roots and the Pythagorean Theorem

Before Class:

☐   Read the objectives on page 459.

☐   Read the **Helpful Hint** box on page 460.

☐   Look over the Pythagorean Theorem and the technique for Finding an Unknown Length of a Right Triangle on page 461.

☐   Complete the exercises:

1.   Evaluate: $3^2 = $ _____

2.   Evaluate: $6^2 + 8^2 = $ _____

3.   Evaluate: $\left(\dfrac{1}{8}\right)^2 = $ _____

4.   Use a calculator to approximate $\sqrt{5}$ to the nearest tenth.

During Class:

☐   **Write your class notes.** Neatly write down **all** examples shown as well as key terms or phrases with definitions. If not applicable or if you were absent, watch the Lecture Series (DVD) for this section and do the same (write down the examples shown as well as key terms or phrases.) Insert more paper as needed.

Class Notes/Example	Your Notes

Answers: **1)** 9   **2)** 100   **3)** $\dfrac{1}{64}$ **4)** ≈ 2.2

163

Section 6.6 Square Roots and the Pythagorean Theorem

Class Notes (continued)	**Your Notes**

(Insert additional paper as needed.)

Section 6.6 Square Roots and the Pythagorean Theorem

Practice:

☐   Complete the Vocabulary and Readiness Check on page 464.

☐   Next, complete any incomplete exercises below. Check and correct your work using the answers and references at the end of this section.

Review this example:	**Your turn:**
1. a. Find $\sqrt{49}$.	**2. a.** Find $\sqrt{121}$.

$\sqrt{49} = \boxed{7}$ because $7^2 = 49$.

Hint: We use $\sqrt{}$ to indicate the positive square root of a positive number.

     b. Find $\sqrt{\dfrac{4}{25}}$.

b. Find $\sqrt{\dfrac{1}{81}}$.

$\sqrt{\dfrac{4}{25}} = \boxed{\dfrac{2}{5}}$ because $\dfrac{2}{5} \cdot \dfrac{2}{5} = \dfrac{4}{25}$.

Review this example:

3. Use a calculator to approximate $\sqrt{80}$ to the nearest thousandth.

$\sqrt{80} \approx \boxed{8.944}$

Your turn:

4. Use a calculator to approximate $\sqrt{15}$ to the nearest thousandth.

Complete this example:

5. Find the length of the hypotenuse of the given right triangle.

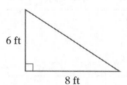

6 ft

8 ft

Use the formula

$$\text{hypotenuse} = \sqrt{(\text{leg})^2 + (\text{other leg})^2}$$

$$= \sqrt{(6)^2 + (8)^2} \quad \text{The legs are 6 feet and 8 feet.}$$

$$= \sqrt{36 + 64}$$

$$= \sqrt{?}$$

$$= \underline{}$$

The length of the hypotenuse is $\boxed{\underline{} \text{ feet.}}$

Your turn:

6. Find the unknown length.

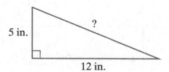

5 in.

?

12 in.

Section 6.6 Square Roots and the Pythagorean Theorem

Review this example:

7. Find the length of the leg in the given right triangle. Give the exact length and a two-decimal-place approximation.

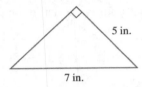

5 in.

7 in.

Use the formula

$$leg = \sqrt{(hypotenuse)^2 - (other\ leg)^2}$$

$$= \sqrt{(7)^2 - (5)^2}$$ The hypotenuse is 7 inches, and the other leg is 5 inches.

$$= \sqrt{49 - 25}$$

$$= \sqrt{24}\ in. \quad \text{Exact answer}$$

$$\approx 4.90\ in. \quad \text{Approximate answer}$$

Your turn:

8. Sketch the right triangle and find the length of the side not given. If necessary, approximate the length to the nearest thousandth.

hypotenuse = 2, leg = 1

	Answer	Text Ref	Video Ref		Answer	Text Ref	Video Ref
1	a. 7 b. $\frac{2}{5}$	a. Ex 1, p. 459 b. Ex 3, p. 459		5	$\sqrt{100} = 10$ 10 feet	Ex 6, p. 461	
2	a. 11 b. $\frac{1}{9}$		a. Sec 6.6, 2/7 b. Sec 6.6, 3/7	6	13 inches		Sec 6.6, 6/7
3	8.944	Ex 4b, p. 460		7	$\sqrt{24}$ inches ≈ 4.90 inches	Ex 8, p. 462	
4	3.873		Sec 6.6, 5/7	8	1.732 units		Sec 6.6, 7/7

☐ **Next, insert your homework.** Make sure you attempt all exercises asked of you and show all work, as in the exercises above. Check your answers if possible. Clearly mark any exercises you were unable to correctly complete so that you may ask questions later. DO NOT ERASE YOUR INCORRECT WORK. THIS IS HOW WE UNDERSTAND AND EXPLAIN TO YOU YOUR ERRORS.

Section 6.7 Congruent and Similar Triangles

Before Class:

☐   Read the objectives on page 468.

☐   Read the ways to determine whether two triangles are congruent on page 468.

☐   Read the definition of similar triangles on page 469.

☐   Complete the exercises. Fill in each blank with **congruent, similar,** or **proportion**.

   1. Two triangles are _____ when they have the same shape but not necessarily the same size.

   2. Two triangles are _____ when they have the same shape and the the same size.

   3. In similar triangles, the measures of corresponding angles are equal and

       corresponding sides are in _____.

During Class:

☐   **Write your class notes.** Neatly write down **all** examples shown as well as key terms or phrases with definitions. If not applicable or if you were absent, watch the Lecture Series (DVD) for this section and do the same (write down the examples shown as well as key terms or phrases). Insert more paper as needed.

Class Notes/Examples	**Your Notes**

Answers: **1)** similar   **2)** congruent   **3)** proportion

Section 6.7 Congruent and Similar Triangles

Class Notes (continued)	**Your Notes**

(Insert additional paper as needed.)

Section 6.7 Congruent and Similar Triangles

Practice:

☐   Complete the Vocabulary and Readiness Check on page 472.

☐   Complete any incomplete exercises below. Check and correct your work using the
        answers and references at the end of this section.

Review these examples:

1. Determine whether triangle *ABC* is congruent
    to triangle *DEF*.

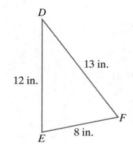

The lengths of all three sides of triangle ABC equal
the lengths of all three sides of triangle DEF.

These two triangles are congruent by Side-
Side-Side.

Your turn:

2. Determine whether the pair of
    triangles is congruent.

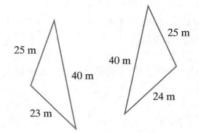

Review this example:

3. Find the ratio of corresponding sides for the
    similar triangles *ABC* and *DEF*.

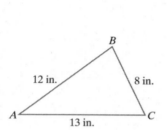

Two corresponding
sides are given.

The ratio of their
lengths is:

$$\frac{12 \text{ feet}}{19 \text{ feet}} = \frac{12}{19}$$

Your turn:

4. Find each ratio of the
    corresponding sides of the given
    similar triangles.

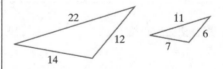

169

Section 6.7 Congruent and Similar Triangles

Review this example:

5. Given that the triangles are similar, find the missing length *n*.

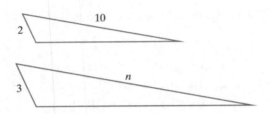

Since the triangles are similar, corresponding sides are in proportion.

The ratio of 2 to 3 is the same as the ratio of 10 to n.

$2 \cdot n = 3 \cdot 10$ Set cross products equal.

$2n = 30$ Multiply.

$\dfrac{2n}{2} = \dfrac{30}{2}$ Divide both sides by 2.

$n = 15$ Simplify.

Your turn:

6. Given that the triangles are similar, find the missing length *n*.

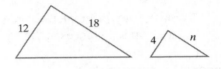

	Answer	Text Ref	Video Ref		Answer	Text Ref	Video Ref
1	congruent SSS	Ex 1, p. 469		**4**	$\dfrac{2}{1}$		Sec 6.7, 2/4
2	not congruent		Sec 6.7, 1/4	**5**	$n = 15$	Ex 3, p. 470	
3	$\dfrac{12}{19}$	Ex 2, p. 470		**6**	$n = 6$		Sec 6.7, 3/4

☐   **Next, insert your homework.** Make sure you attempt all exercises asked of you and show all work, as in the exercises above. Check your answers if possible. Clearly mark any exercises you were unable to correctly complete so that you may ask questions later. DO NOT ERASE YOUR INCORRECT WORK. THIS IS HOW WE UNDERSTAND AND EXPLAIN TO YOU YOUR ERRORS.

Preparing for the Chapter 6 Test

Start preparing for your Chapter 6 Test as soon as possible. Pay careful attention to any instructor discussion about this test, especially discussion on what sections you will be responsible for, etc.

☐ Work the Chapter 6 Vocabulary Check on page 477.

☐ Read both columns (Definitions and Concepts, and Examples) of the Chapter 6 Highlights starting on page 477.

☐ Read your Class Notes/Examples for each section covered on your Chapter 6 Test. Look for any unresolved questions you may have.

☐ Complete as many of the Chapter 6 Review exercises as possible (pages 482 - 486). Remember, the odd answers are in the back of your text.

☐ **Most important:** Place yourself in "test" conditions (see below) and work the Chapter 6 Test (pages 489 – 490) as a practice test the day before your actual test. To honestly assess how you are doing, try the following:

- Work on a few blank sheets of paper.
- Give yourself the same amount of time you will be given for your actual test.
- Complete this Chapter 6 Practice Test without using your notes or your text.
- If you have any time left after completing this practice test, check your work and try to find any errors on your own.
- Once done, use the back of your book to check ALL answers.
- Try to correct any errors on your own.
- Use the Chapter Test Prep Video (CTPV) to correct any errors you were unable to correct on your own. You can find these videos in the Interactive DVD Lecture Series, in MyMathLab, and on YouTube. Search MartinGayBasicMath and click channels.

I wish you the best of luck….Elayn Martin-Gay

Section 7.1 Reading Pictographs, Bar Graphs, Histograms, and Line Graphs

Before Class:

☐   Read the objectives on page 494.

☐   Read the paragraphs below each of the following objectives: Objective A page 494, Objective B page 495, Objective C page 498, and Objective D page 500.

☐   Complete the exercises:

1. Give one advantage and one disadvantage for using a pictograph to display information.

2. What does the ⟩ symbol alert us to when seen on a vertical scale of a graph? (p. 498)

3. The bars in a histogram lie side by side with no _____ between them.

4. Give two advantages for using a line graph to display information. (p. 500)

During Class:

☐   **Write your class notes.** Neatly write down **all** examples shown as well as key terms or phrases with definitions. If not applicable or if you were absent, watch the Lecture Series (DVD) for this section and do the same (write down the examples shown as well as key terms or phrases). Insert more paper as needed.

Class Notes/Examples	**Your Notes**

Answers: **1)** Pictographs allow comparisons to be easily made. It is hard to tell what fractional part of a symbol is shown. **2)** Numbers are missing on that scale. **3)** space **4)** Line graphs can be used to visualize relationships between two quantities and can be useful in showing a change over time.

Section 7.1 Reading Pictographs, Bar Graphs, Histograms, and Line Graphs

Class Notes (continued)	**Your Notes**

(Insert additional paper as needed.)

Section 7.1 Reading Pictographs, Bar Graphs, Histograms, and Line Graphs

Practice:

☐   Complete the Vocabulary and Readiness Check on page 502.

☐   Next, complete any incomplete exercises below. Check and correct your work using the answers and references at the end of this section.

Complete this example:

1. Use the information given to draw a vertical bar graph.

Average Caffeine Content of Selected Foods			
Food	**Milligrams**	**Food**	**Milligrams**
Brewed coffee (percolator, 8 ounces)	124	Instant coffee (8 ounces)	104
Brewed decaffeinated coffee (8 ounces)	3	Brewed tea (U.S. brands, 8 ounces)	64
Coca-Cola Classic (8 ounces)	31	Mr. Pibb (8 ounces)	27
Dark chocolate (semisweet, $1\frac{1}{2}$ ounces)	30	Milk chocolate (8 ounces)	9

(*Sources:* International Food Information Council and the Coca-Cola Company)

The milligrams scale starts at ____ and then stops at

____. The scale shows multiples of ____.

Draw in all vertical bars using the above data.

Your turn:

2. Use the information given to draw a vertical bar graph.

Fiber Content of Selected Foods	
Food	**Grams of Total Fiber**
Kidney beans $\left(\frac{1}{2}c\right)$	4.5
Oatmeal, cooked $\left(\frac{3}{4}c\right)$	3.0
Peanut butter, chunky (2 tbsp)	1.5
Popcorn (1 c)	1.0
Potato, baked with skin (1 med)	4.0
Whole wheat bread (1 slice)	2.5

(*Sources:* American Dietetic Association and National Center for Nutrition and Dietetics)

Section 7.1 Reading Pictographs, Bar Graphs, Histograms, and Line Graphs

Review this example:

3. Use the graph below to answer the following questions:

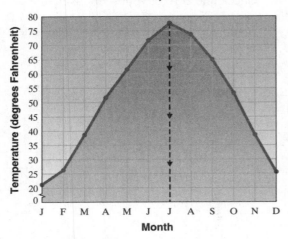

Average Daily Temperature for Omaha, Nebraska

Source: National Climatic Data Center

   a. During what month is the average daily temperature the highest?

   b. During what month, from July through December, is the average daily temperature 65°F?

a. The month with the highest temperature is marked with the red dot. This month is (July.)

b. Find the 65°F mark on the vertical temperature scale and move to the right until a point on the right side of the graph is reached. This month is (September.)

Your turn:

4. The following line graph shows the total points scored by both teams in the NFL Super Bowl from 2003 through 2009.

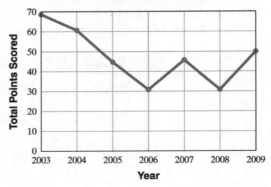

Total Points Scored in Super Bowl

a. Find the total points scored in the Super Bowl in 2003.

b. During which of the years shown were the total points scored in the Super Bowl greater than 60?

c. During which year(s) shown was the total score in the Super Bowl the lowest?

	Answer	Text Ref	Video Ref		Answer	Text Ref	Video Ref
1	0, 140; multiples of 20	Ex 4, p. 497		3	a. July b. September	Ex 9, p. 500	
2	Ask instructor		Sec 7.1, 5/13	4	a. 69 b. 2003, 2004 c. 2006, 2008		Sec 7.1, 10, 12-13/13

☐   **Next, insert your homework.** Make sure you attempt all exercises asked of you and show all work, as in the exercises above. Check your answers if possible. Clearly mark any exercises you were unable to correctly complete so that you may ask questions later. DO NOT ERASE YOUR INCORRECT WORK. THIS IS HOW WE UNDERSTAND AND EXPLAIN TO YOU YOUR ERRORS.

Section 7.2 Reading Circle Graphs

Before Class:

☐   Read the objectives on page 508.

☐   Read the **Helpful Hint** boxes on pages 509 and 510.

☐   Complete the exercises:

1.   Since a circle graph represents a whole, the percents should add to _____ .

2.   Write the ratio of 12 people to 16 people as a fraction in simplest form.

3.   Write the ratio of 65 books to 100 books as a fraction in simplest form.

4.   What is 20% of 40?

During Class:

☐   **Write your class notes.** Neatly write down **all** examples shown as well as key terms or phrases with definitions. If not applicable or if you were absent, watch the Lecture Series (DVD) for this section and do the same (write down the examples shown as well as key terms or phrases.) Insert more paper as needed.

Class Notes/Examples	**Your Notes**

Answers: **1)**   100% or 1   **2)** $\dfrac{3}{4}$ **3)** $\dfrac{13}{20}$ **4)** 8

Section 7.2 Reading Circle Graphs

Class Notes (continued)	Your Notes

(Insert additional paper as needed.)

Section 7.2 Reading Circle Graphs

Practice:

☐   Complete the Vocabulary and Readiness Check on page 512.

☐   Next, complete any incomplete exercises below. Check and correct your work using the answers and references at the end of this section.

Review this example:

1. This particular graph shows the favorite sport for 100 adults. Find the ratio of adults preferring basketball to total adults. Write the ratio as a fraction in simplest form.

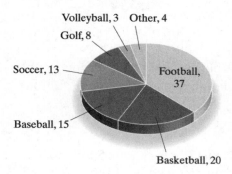

The ratio is:

$$\frac{\text{people preferring basketball}}{\text{total adults}} = \frac{20}{100}$$

$$= \left(\frac{1}{5}\right)$$

Your turn:

2. 700 college students were asked where they live while attending college.

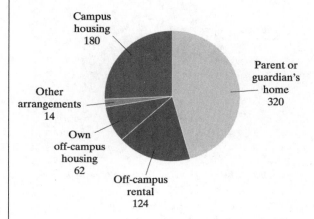

a. Where do most of these college students live?

b. Find the ratio of students living in campus housing to total students.

179

Section 7.2 Reading Circle Graphs

Review this example:

3. The following graph shows the percent of visitors to the United States in a recent year by various regions. Determine the percent of visitors who came to the United States from Mexico and Canada.

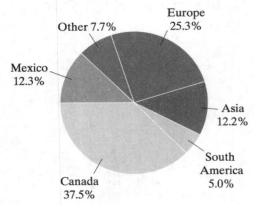

Visitors to U.S. by Region

Source: Office of Travel and Tourism Industries

$12.3\% + 37.5\% = \boxed{49.8\%}$

Your turn:

4. The following circle graph shows the percent of the types of books available at Midway Memorial Library.

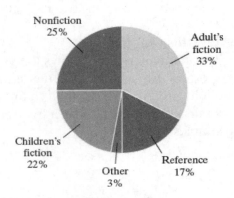

a. What percent of books are classified as some type of fiction?

b. If this library has 125,600 books, how many books are Nonfiction?

	Answer	Text Ref	Video Ref		Answer	Text Ref	Video Ref
1	$\dfrac{1}{5}$	Ex 1, p. 508		**3**	49.8%	Ex 2, p. 508	
2	a. Parent or guardian's home b. $\dfrac{9}{35}$		Sec 7.2, 1-2/6	**4**	a. 55% b. 31,400		Sec 7.2, 3/6 Sec 7.2, 5/6

☐ **Next, insert your homework.** Make sure you attempt all exercises asked of you and show all work, as in the exercises above. Check your answers if possible. Clearly mark any exercises you were unable to correctly complete so that you may ask questions later. DO NOT ERASE YOUR INCORRECT WORK. THIS IS HOW WE UNDERSTAND AND EXPLAIN TO YOU YOUR ERRORS.

Section 7.3 Mean, Median, and Mode

Before Class:

☐ Read the objectives on page 518.

☐ Read the **Helpful Hint** boxes on page 520.

☐ Read the following definitions: **mean** page 518, **median** page 519, and **mode** page 520.

☐ Complete the exercises:

1. Find the average of the two exam scores: 70, 80

2. Find the average of the three exam scores: 70, 80, 90

3. Find the average of the four exam scores: 70, 80, 90, 100

During Class:

☐ **Write your class notes.** Neatly write down **all** examples shown as well as key terms or phrases with definitions. If not applicable or if you were absent, watch the Lecture Series (DVD) for this section and do the same (write down the examples shown as well as key terms or phrases.) Insert more paper as needed.

Class Notes/Example	**Your Notes**

Answers: **1)** 75   **2)** 80   **3)** 85

Section 7.3 Mean, Median, and Mode

Class Notes (continued)	**Your Notes**

(Insert additional paper as needed.)

Section 7.3 Mean, Median, and Mode

Practice:

☐   Complete the Vocabulary and Readiness Check on page 521.

☐   Next, complete any incomplete exercises below. Check and correct your work using the answers and references at the end of this section.

Review this example:

1. Determine the following for each set of numbers.

   a. Find the *mean*:

     13.2, 11.8, 10.7, 16.2, 15.9, 13.8, 18.5

To find the mean (or average), find the sum of the numbers and divide by 7 (the number of items).

$$\text{Mean} = \frac{13.2 + 11.8 + 10.7 + 16.2 + 15.9 + 13.8 + 18.5}{7}$$

$$\text{Mean} = \frac{100.1}{7} = \boxed{14.3}$$

   b. Find the *median*:

     25, 54, 56, 57, 60, 71, 98

This list is in numerical order. The number in the middle is the median.

     The median is 57.

   c. Find the *mode*:

     11, 14, 14, 16, 31, 56, 65, 77, 77, 78, 79

There are two numbers that occur the most frequently. This list of numbers has two modes.

     The two modes are 14 and 77.

Your turn:

2. For the set of numbers:

   7.6, 8.2, 8.2, 9.6, 5.7, 9.1

   mean

   median

Helpful Hint Don't forget to write the numbers in order from smallest to largest before finding the median.

   mode

Section 7.3 Mean, Median, and Mode

Review this example:

3. In the chart below the grades are given for a student for a particular semester. Find the grade point average. If necessary, round to the nearest hundredth.

Course	Grade	Credit Hours
College mathematics	A	3
Biology	B	3
English	A	3
PE	C	1
Social studies	D	2

The point values for the different possible grades are given below.

A: 4,   B: 3,   C: 2,   D: 1,   F: 0

To find the GPA multiply each credit hour by the point value of the grade for that class. The GPA will be the sum of these products divided by the sum of the credit hours. Note: The sum of the credit hours is 12.

$$\text{GPA} = \frac{3(4) + 3(3) + 3(4) + 1(2) + 2(1)}{12} = \frac{37}{12} \approx 3.08$$

GPA ≈ 3.08

Your turn:

4. In the chart below the grades are given for a student for a particular semester. Find the grade point average. If necessary, round to the nearest hundredth.

Grade	Credit Hours
B	3
C	3
A	4
C	4

	Answer	Text Ref	Video Ref		Answer	Text Ref	Video Ref
1	Mean: 14.3 Median: 57 Mode: 14,77	Ex 1b, p. 518 Ex 3, p. 519 Ex 5, p. 520		**3**	$\frac{37}{12} \approx 3.08$	Ex 2, p. 519	
2	Mean: 8.1 Median: 8.2 Mode: 8.2		Sec 7.3, 1,3-4/4	**4**	2.79		Sec 7.3, 2/4

☐   **Next, insert your homework.** Make sure you attempt all exercises asked of you and show all work, as in the exercises above. Check your answers if possible. Clearly mark any exercises you were unable to correctly complete so that you may ask questions later. DO NOT ERASE YOUR INCORRECT WORK. THIS IS HOW WE UNDERSTAND AND EXPLAIN TO YOU YOUR ERRORS.

Section 7.4 Counting and Introduction to Probability

Before Class:

☐   Read the objectives on page 523.

☐   Read the **Probability of an Event** box on page 524.

☐   Complete the exercises: The following chart shows all the ways for rolling a pair of dice.
        Look for a pattern and complete the chart.

Outcome
for rolling
1 die

	1	2	3	4	5	6
1	(1,1) sum = 2	(1,2) sum = 3	(1,3) sum = 4	(1,4) sum = 5	(1,5) sum = 6	(1,6) sum = 7
2	(2,1) sum = 3	(2,2) sum = 4	(2,3) sum = 5	(2,4) sum = 6		
3	(3,1) sum = 4	(3,2) sum = 5				
4	(4,1) sum = 5					
5						
6						

During Class:

☐   **Write your class notes.** Neatly write down **all** examples shown as well as key terms or
        phrases with definitions. If not applicable or if you were absent, watch the Lecture Series
        (DVD) for this section and do the same (write down the examples shown as well as key terms
        or phrases). Insert more paper as needed.

Class Notes/Examples	**Your Notes**

Answers: **1)** patterns: sums along diagonals are the same, columns down increase by 1

Section 7.4 Counting and Introduction to Probability

Class Notes (continued)	Your Notes

(Insert additional paper as needed.)

Section 7.4 Counting and Introduction to Probability

Practice:

☐   Complete the Vocabulary and Readiness Check on page 526.

☐   Complete any incomplete exercises below. Check and correct your work using the answers and references at the end of this section.

Review this example:	**Your turn:**
1. Draw a tree diagram for the experiment: tossing a coin twice. Find the total number of outcomes.	**2.** Draw a tree diagram for the experiment: choosing a letter in the word MATH, then a number (1, 2, or 3). Find the total number of outcomes.

The possible outcomes for flipping a coin are heads (H) and tails (T).

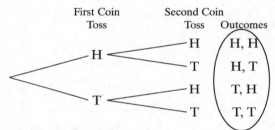

First Coin Toss   Second Coin Toss   Outcomes

H, H
H, T
T, H
T, T

There are ④ possible outcomes.

Review this example:

3. If a die is rolled one time, find the probability of rolling a 3 or 4.

There are 6 possible outcomes when rolling a die.

Possible outcomes: 1 , 2 ,③ ④, 5 , 6

Probability of a 3 or 4:

$= \dfrac{2}{6}$ ⟵ Number of ways the event can occur
       ⟵ Number of possible outcomes

 Simplest Form

Your turn:

4. If a single die is tossed once, find the probability of each event.

a.   rolling a 5

b.   rolling an even number

Section 7.4 Counting and Introduction to Probability

Review this example:

5. Find the probability of choosing a red marble from a box containing 1 red, 1 yellow, and 2 blue marbles.

1 way that event can occur

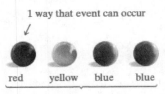

red   yellow   blue   blue

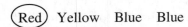

4 possible outcomes

 Yellow   Blue   Blue

Probability of getting a red marble:

$$= \left(\frac{1}{4}\right) \begin{array}{l} \leftarrow \text{ Number of ways the event can occur} \\ \leftarrow \text{ Number of possible outcomes} \end{array}$$

Your turn:

6. Suppose the spinner shown is spun once. Find the probability of the following event:

The result of the spin is 2.

	Answer	Text Ref	Video Ref		Answer	Text Ref	Video Ref
1	4 possible outcomes	Ex 1, p. 523		**4**	a. $\frac{1}{6}$ b. $\frac{1}{2}$		Sec 7.4, 2-3/4
2	12 outcomes		Sec 7.4, 1/4	**5**	$\frac{1}{4}$	Ex 5, p. 525	
3	$\frac{1}{3}$	Ex 4, p. 525		**6**	$\frac{1}{3}$		Sec 7.4, 4/4

☐   **Next, insert your homework.** Make sure you attempt all exercises asked of you and show all work, as in the exercises above. Check your answers if possible. Clearly mark any exercises you were unable to correctly complete so that you may ask questions later. DO NOT ERASE YOUR INCORRECT WORK. THIS IS HOW WE UNDERSTAND AND EXPLAIN TO YOU YOUR ERRORS.

Preparing for the Chapter 7 Test

Start preparing for your Chapter 7 Test as soon as possible. Pay careful attention to any instructor discussion about this test, especially discussion on what sections you will be responsible for, etc.

☐   Work the Chapter 7 Vocabulary Check on page 531.

☐   Read both columns (Definitions and Concepts, and Examples) of the Chapter 7 Highlights starting on page 531.

☐   Read your Class Notes/Examples for each section covered on your Chapter 7 Test. Look for any unresolved questions you may have.

☐   Complete as many of the Chapter 7 Review exercises as possible (starting on page 534). Remember, the odd answers are in the back of your text.

☐   **Most important:** Place yourself in "test" conditions (see below) and work the Chapter 7 Test (pages 539-542) as a practice test the day before your actual test. To honestly assess how you are doing, try the following:

- Work on a few blank sheets of paper.
- Give yourself the same amount of time you will be given for your actual test.
- Complete this Chapter 7 Practice Test without using your notes or your text.
- If you have any time left after completing this practice test, check your work and try to find any errors on your own.
- Once done, use the back of your book to check ALL answers.
- Try to correct any errors on your own.
- Use the Chapter Test Prep Video (CTPV) to correct any errors you were unable to correct on your own. You can find these videos in the Interactive DVD Lecture Series, in MyMathLab, and on YouTube. Search MartinGayDevMath and click "Channels."

I wish you the best of luck….Elayn Martin-Gay

Section 8.1 Symbols and Sets of Numbers

Before Class:

☐   Read the objectives on page 547.

☐   Read the Equality and Inequality Symbols box on page 547 and the Order Property for Real Numbers box on page 548.

☐   Read the **Helpful Hint** boxes on page 547, 548, 549, 550 and 553.

☐   Complete the exercises:

    1.   Write each statement using an equality or inequality symbol.

       a.   2 is less than 5   _____

       b.   5 is greater than 2   _____

       c.   12 is equal to $3 \cdot 4$ _____

       d.   6 is greater than or equal to 1   _____

    2.   A common name for rational numbers is _____. (see page 550)

During Class:

☐   **Write your class notes.** Neatly write down **all** examples shown as well as key terms or phrases with definitions. If not applicable or if you were absent, watch the Lecture Series (DVD) for this section and do the same (write down the examples shown as well as key terms or phrases). Insert more paper as needed.

Class Notes/Examples	**Your Notes**

Answers: **1)** a. $2 < 5$ b. $5 > 2$ c. $12 = 3 \cdot 4$ d. $6 \geq 1$ **2)** fractions

191

Section 8.1 Symbols and Sets of Numbers

Class Notes (continued)

Your Notes

(Insert additional paper as needed.)

Section 8.1 Symbols and Sets of Numbers

Practice:

☐   Complete the Vocabulary and Readiness Check on page 554.

☐   Next, complete any incomplete exercises below. Check and correct your work using the
      answers and references at the end of this section.

Review this example:	**Your turn:**
1. Insert $<$, $>$, or $=$ in the space between the paired numbers to make each statement true.	2. Insert $<$, $>$, or $=$ in the space between the paired numbers to make each statement true.

    a. -5 -6 b. 3.195   3.2

   a. -5 is to the right of -6 on the number line.

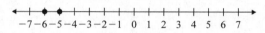

    Therefore, $\boxed{-5 > -6}$.

   b. Compare digits in the same place value.
    The ones digits are equal. The tenths
    digits are different: $0.1 < 0.2$.

    Thus, $\boxed{3.195 < 3.2}$.

a. 7   3

b. 0   7

Review this example:	**Your turn:**
3. Determine whether the statement is true or false: $2 < 3$	4. Is the statement $11 \le 11$ true or false?

   $2 < 3$ $\boxed{\text{True.}}$ Since 2 is left of 3 on a
                   number line.

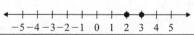

Review this example:	**Your turn:**
5. Write the sentence as a mathematical statement:	6. Write each sentence as a mathematical statement.

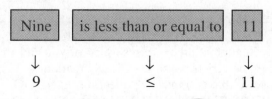

        ↓   ↓   ↓
        9             $\le$ 11

The mathematical statement is $\boxed{9 \le 11}$.

a. Five is greater than or equal to four.

b. Fifteen is not equal to negative two.

193

Section 8.1 Symbols and Sets of Numbers

Review this example:

7. Given the set $\left\{-2, 0, \dfrac{1}{4}, 112, -3, 11, \sqrt{2}\right\}$, list

the numbers in the set that belong to the set of:

   a.   Whole Numbers   b.   Irrational numbers

a. The whole numbers are ⟨0, 11 and 112⟩.

b. The irrational number is ⟨$\sqrt{2}$⟩.

Your turn:

8. Tell which set or sets each number belongs to: natural numbers, whole numbers, integers, rational numbers, irrational numbers, or real numbers.

   a. 0   b. $\dfrac{2}{3}$

Review this example:

9. Insert <, >, or = in the space between the paired numbers to make each statement true.

   a. $|-3|$ $|-2|$ b. $|-9|$ $|-9.7|$

a. $|-3| = 3$ and $|-2| = 2$. Since $3 > 2$, ⟨$|-3| > |-2|$⟩

b. $|-9| = 9$ and $|-9.7| = 9.7$.

Since $9 < 9.7$, ⟨$|-9| < |-9.7|$⟩.

Your turn:

10. Insert <, >, or = in the space between the paired numbers to make each statement true.

   a. $|-5|$ -4

   b. $|0|$ $|-8|$

	Answer	Text Ref	Video Ref		Answer	Text Ref	Video Ref
1	a. > b. <	Ex 10a,10b, p. 552		6	a. $5 \geq 4$ b. $15 \neq -2$		Sec 8.1, 4-5/14
2	a. > b. <		Sec 8.1, 1-2/14	7	a. $0, 11, 112$ b. $\sqrt{2}$	Ex 11b, 11e, p. 552	
3	True	Ex 1, p. 548		8	a. whole, integer, rational, real b. rational, real		Sec 8.1, 6-7/14
4	True $11 = 11$		Sec 8.1, 3/14	9	a. > b. <	Ex 13c, 13d, p. 553	
5	$9 \leq 11$	Ex 7a, p. 548		10	a. > b. <		Sec 8.1, 13-14/14

☐   **Next, insert your homework.** Make sure you attempt all exercises asked of you and show all work, as in the exercises above. Check your answers if possible. Clearly mark any exercises you were unable to correctly complete so that you may ask questions later. DO NOT ERASE YOUR INCORRECT WORK. THIS IS HOW WE UNDERSTAND AND EXPLAIN TO YOU YOUR ERRORS.

Section 8.2 Exponents, Order of Operations, and Variable Expressions

Before Class:

☐ Read the objectives on page 559.

☐ Read the **Helpful Hint** boxes on pages 559, 560, 561, 562, 563 and 564.

☐ Read the Order of Operations box on page 560 and the key words and phrases box on page 563.

☐ Complete the exercises:

1. Simplify.

a. $2 \cdot 2 \cdot 2 \cdot 2 \cdot 2$ b. $10 \cdot 10 \cdot 10$ c. $\left(\dfrac{2}{3}\right)\left(\dfrac{2}{3}\right)$ d. $(0.3)(0.3)$

2. Simplify.

a. $|20 - 12|$ b. $|-8.5|$

3. Is $6x - 7$ an equation or an expression? (p. 562)

During Class:

☐ **Write your class notes.** Neatly write down **all** examples shown as well as key terms or phrases with definitions. If not applicable or if you were absent, watch the Lecture Series (DVD) for this section and do the same (write down the examples shown as well as key terms or phrases). Insert more paper as needed.

Class Notes/Examples | **Your Notes**

Answers: **1)** a. 32   b. 1000   c. $\dfrac{4}{9}$ d. 0.09   **2)** a. 8   b. 8.5   **3)** expression

Section 8.2 Exponents, Order of Operations, and Variable Expressions

Class Notes (continued)	**Your Notes**

(Insert additional paper as needed.)

Section 8.2 Exponents, Order of Operations, and Variable Expressions

Practice:

☐   Complete the Vocabulary and Readiness Check on page 566.

☐   Next, complete any incomplete exercises below. Check and correct your work using the answers and references at the end of this section.

Complete this example:	**Your turn:**
1. Evaluate.	**2.** Evaluate.

a. 5^3 b. $\left(\dfrac{3}{7}\right)^2$ a. 3^3 b. $\left(\dfrac{2}{3}\right)^4$

a. 5^3 is read as 5 to the third power.

$5^3 = 5 \cdot 5 \cdot 5 = 25 \cdot 5 = \boxed{125}$

b. $\left(\dfrac{3}{7}\right)^2$ is read as $\dfrac{3}{7}$ to the second power.

$\left(\dfrac{3}{7}\right)^2 = \dfrac{3}{7} \cdot \dfrac{3}{7} = \bigcirc$

Review this example:

3. Simplify each expression.

a. $20 \div 5 \cdot 4$ b. $1 + 2\left[5(2 \cdot 3 + 1) - 10\right]$

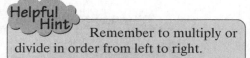

Helpful Hint
Remember to multiply or divide in order from left to right.

a. $20 \div 5 \cdot 4 = 4 \cdot 4$ Divide $20 \div 5$.

$= \boxed{16}$ Multiply $4 \cdot 4$.

b. $1 + 2\left[5(2 \cdot 3 + 1) - 10\right]$ Simplify innermost set of parentheses.

$1 + 2\left[5(2 \cdot 3 + 1) - 10\right] = 1 + 2[5(7) - 10]$

$= 1 + 2[35 - 10]$ Multiply $5(7)$.

$= 1 + 2[25]$ Subtract.

$= 1 + 50$ Multiply $2(25)$.

$= \boxed{51}$ Add.

Your turn:

4. Simplify each expression.

a. $5 + 6 \cdot 2$

b. $2\left[5 + 2(8 - 3)\right]$

c. $\dfrac{|6 - 2| + 3}{8 + 2 \cdot 5}$

197

Section 8.2 Exponents, Order of Operations, and Variable Expressions

Review this example:	**Your turn:**
5. Evaluate the expression $x^2 - y^2$ when $x = 3$ and $y = 2$. Replace x with 3 and y with 2. Then, simplify. $x^2 - y^2 = (3)^2 - (2)^2 = 9 - 4 = \boxed{5}$	**6.** Evaluate the expression $\lvert 2x + 3y \rvert$ when $x = 1$ and $y = 3$.

Review this example:	**Your turn:**
7. Decide whether the number 2 is a solution to the equation $3x + 10 = 8x$. $3(2) + 10 \overset{?}{=} 8(2)$ Replace x with 2. $6 + 10 \overset{?}{=} 16$ Simplify each side. $16 = 16$ True. Yes, 2 is a solution.	**8.** Decide whether the number 6 is a solution to the equation $3x - 10 = 8$.

Review this example:	**Your turn:**
9. Write the phrase as an algebraic expression. Let x represent the unknown number. *Five times a number, increased by 7* $5x + 7$ Times means multiply and increased by means to add.	**10.** Write the phrase as an algebraic expression. Let x represent the unknown number. *Three times a number, increased by 22*

	Answer	Text Ref	Video Ref		Answer	Text Ref	Video Ref
1	a. 125 b. $\dfrac{9}{49}$	Ex 1b, 1e, p. 559		6	11		Sec 8.2, 6/12
2	a. 27 b. $\dfrac{16}{81}$		Sec 8.2, 1-2/12	7	16 = 16 solution	Ex 8, p. 563	
3	a. 16 b. 51	Ex 3, 5, p. 560		8	solution		Sec 8.2, 8/12
4	a. 17 b. 30 c $= \dfrac{7}{18}$		Sec 8.2, 3-5/12	9	$5x + 7$	Ex 9e p. 563	
5	5	Ex 7e p. 562		10	$3x + 22$		Sec 8.2, 10/12

☐   **Next, insert your homework.** Make sure you attempt all exercises asked of you and show all work, as in the exercises above. Check your answers if possible. Clearly mark any exercises you were unable to correctly complete so that you may ask questions later. DO NOT ERASE YOUR INCORRECT WORK. THIS IS HOW WE UNDERSTAND AND EXPLAIN TO YOU YOUR ERRORS.

Section 8.3 Adding Real Numbers

Before Class:

☐　　Read the objectives on page 570.

☐　　Read the Adding Real Numbers box on page 571 and Additive Inverse box on page 572.

☐　　Complete the exercises:

1. Simplify:

   a.　$|8| =$ _____　　b.　$|-8| =$ _____　　c.　$|-4.6| =$ _____　　d.　$\left|-\dfrac{2}{3}\right| =$ _____

2. Add or Subtract as Indicated:

   a.　$3.5 + 2.7$　　b.　$3.5 - 2.7$　　c.　$\dfrac{4}{5} + \dfrac{2}{3}$　　d.　$\dfrac{4}{5} - \dfrac{2}{3}$

3. What is the additive inverse of 3? (p. 572)

4. What is the sum of $3 + (-3)$? (p. 573)

During Class:

☐　　**Write your class notes.**　Neatly write down **all** examples shown as well as key terms or phrases with definitions. If not applicable or if you were absent, watch the Lecture Series (DVD) for this section and do the same (write down the examples shown as well as key terms or phrases). Insert more paper as needed.

Class Notes/Examples	**Your Notes**

Answers:

1) a. 8　 b. -8　 c. 4.6　 d. $\dfrac{2}{3}$　 **2)** a. 6.2　 b. 0.8　 c. $\dfrac{22}{15}$　 d. $\dfrac{2}{15}$　 **3)** -3　 **4)** 0

Section 8.3 Adding Real Numbers

Class Notes (continued)	Your Notes

(Insert additional paper as needed.)

Practice:

☐   Complete the Vocabulary and Readiness Check on page 576.

☐   Next, complete any incomplete exercises below. Check and correct your work using the answers and references at the end of this section.

Review this example:	**Your turn:**

Review this example:

1. Add: $(-7)+(-6)$

Adding two numbers with the SAME SIGN:

Step 1: Add their absolute values.

$$7+6=13$$

Step 2: Their common sign is negative, so the sum is negative.

$$-7+(-6)=\boxed{-13}$$

Your turn:

2. a. Add: $-2+(-3)$

b. Add: $-\dfrac{7}{10}+\left(-\dfrac{3}{5}\right)$

Review this example:

3. a. Add: $-2+10$

Adding two numbers with DIFFERENT SIGNS:

Step 1: Find the larger absolute value minus the smaller absolute value.

$$10-2=8$$

Step 2: 10 has the larger absolute value and its sign is positive, so the sum is positive.

$$-2+10=\boxed{8}$$

    b. Add: $(-10)+4$

Adding two numbers with DIFFERENT SIGNS:

Step 1: Find the larger absolute value minus the smaller absolute value.

$$10-4=6$$

Step 2: -10 has the larger absolute value and its sign is negative, so the sum is negative.

$$(-10)+4=\boxed{-6}$$

Your turn:

4. a. Add: $-5+9$

b. Add: $5+-7$

c. Add: $6.3+(-8.4)$

201

Section 8.3 Adding Real Numbers

Complete this example:	**Your turn:**
5. Add $3+(-7)+(-8)$.	**6.** Add: $6+(-4)+9$

$= 3+(-7)+(-8)$ Add the numbers from left

$= -4+(-8)$ to right.

$= \bigcirc$

Review this example:	**Your turn:**
7. Evaluate $2x+y$ for $x=3$ and $y=-5$.	**8.** Evaluate $3x+y$ for $x=2$, $y=-3$.

$2x+y = 2\cdot 3+(-5)$ Replace x with 3 and y

$= 6+(-5)$ with -5.

$= \boxed{1}$

Review this example:	**Your turn:**
9. Simplify. a. $-(-10)$ b. $-|-6|$	**10.** Simplify. $|-6|+(-61)$

a. $-(-10) = \boxed{10}$ The opposite of -10 is 10.

b. $-|-6| = \boxed{-6}$ The opposite of the absolute value of -6 is the opposite of 6, or -6.

	Answer	Text Ref	Video Ref		Answer	Text Ref	Video Ref
1	-13	Ex 4, p. 571		6	11		Sec 8.3, 8/11
2	$a.-5,$ $b.-\dfrac{13}{10}$		a. Sec 8.3, 2/11 b. not a video p. 576 #37	7	1	Ex 20, p. 574	
3	$a.\,8, b.-6$	Ex 7, p. 572 Ex 5, p. 571		8	3		Sec 8.3, 10/11
4	$a.\,4$ $b.-2$ $c.-2.3$		a. not a video p. 576 #14 b-c. Sec 8.3, 5/11, 6/11	9	$a.\,10$ $b.-6$	Ex 17a,d p. 573	
5	-12	Ex 12a, p. 572		10	-55		Sec 8.3, 7/11

☐   **Next, insert your homework.** Make sure you attempt all exercises asked of you and show all work, as in the exercises above. Check your answers if possible. Clearly mark any exercises you were unable to correctly complete so that you may ask questions later. DO NOT ERASE YOUR INCORRECT WORK. THIS IS HOW WE UNDERSTAND AND EXPLAIN TO YOU YOUR ERRORS.

Section 8.4 Subtracting Real Numbers

Before Class:

☐ Read the objectives on page 580.

☐ Read the **Helpful Hint** boxes on pages 580 and 582.

☐ Read the definition of Complementary and Supplementary Angles on page 583.

☐ Complete the exercises:

1. $-6 + (-8) = $ ____

2. $6 + (-2) = $ ____

3. $(-6) + (-2) = $ ____

4. $(-6) + (-2) + (-4) = $ ____

5. Evaluate $x + 3y$ for $x = -7$ and $y = 2$.

6. Determine if the sum of a positive number and a negative number always results in a negative number. If false, provide a non-example.

During Class:

☐ **Write your class notes.** Neatly write down **all** examples shown as well as key terms or phrases with definitions. If not applicable or if you were absent, watch the Lecture Series (DVD) for this section and do the same (write down the examples shown as well as key terms or phrases). Insert more paper as needed.

Class Notes/Examples	**Your Notes**

Answers: **1)** -14　**2)** 4　**3)** -8　**4)** -12　**5)** $-7 + 3(2) = -7 + 6 = -1$　**6)** False, because $7 + (-3) = 4$ (non-examples will vary).

Section 8.4 Subtracting Real Numbers

Class Notes (continued)

(Insert additional paper as needed.)

Practice:

☐   Complete the Vocabulary and Readiness Check on page 585.

☐   Next, complete any incomplete exercises below. Check and correct your work using the answers and references at the end of this section.

Review this example:
1. Subtract the following numbers.

Add the first number to the opposite of the second number.

   a. $-13-4=-13+(-4)=\boxed{-17}$

   b. $5.3-(-4.6)=5.3+4.6=\boxed{9.9}$

   c. $-1-(-7)=-1+7=\boxed{6}$

   d. $-\dfrac{3}{10}-\dfrac{5}{10}=-\dfrac{3}{10}+\left(-\dfrac{5}{10}\right)=-\dfrac{8}{10}=\boxed{-\dfrac{4}{5}}$

Your turn:
2. Subtract.

   a. $16-(-3)$

   b. $-\dfrac{3}{11}-\left(-\dfrac{5}{11}\right)$

   c. $8.3-11.2$

Complete this example:
3. Subtract 8 *from* -4.

$-4-8=-4+(-8)$

$=\bigcirc$

Your turn:
4. Subtract 9 *from* -4.

Review this example:
5. Simplify: $-14-8+10-(-6)$

You will rewrite all differences as equivalent sums, and then simplify from left to right.

$=-14-8+10-(-6)$
$=-14+(-8)+10+6$
$=-22+10+6$
$=-12+6$
$=\boxed{-6}$

Your turn:
6. Simplify: $-10-(-8)+(-4)-20$

Section 8.4 Subtracting Real Numbers

Review this example:	**Your turn:**
7. Evaluate $x^2 - y$ for $x = 2$ and $y = -5$. Replace x with 2 and y with -5 in $x^2 - y$. $x^2 - y = (2)^2 - (-5)$ $= 4 - (-5) = 4 + 5 = 9$	**8.** Evaluate $\dfrac{9-x}{y+6}$ for the given replacement values. $x = -5$ and $y = 4$

Complete this example:	**Your turn:**
9. Find the measure of the unknown angle x. 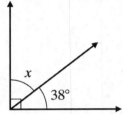 These angles are complementary. The sum of complementary angles is _____. $m\angle x = $ _____ $ - 38° = $ _____	**10.** Find x if the angles below are complementary angles. 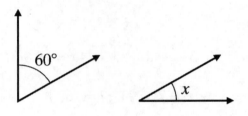

	Answer	Text Ref	Video Ref		Answer	Text Ref	Video Ref
1	a. -17 b. 9.9 c. 11 d. $-\dfrac{4}{5}$	a. Ex 1a, p. 580 b. Ex 2, p. 580 c. Ex 1b, p. 580 d. Ex 3, p. 580		**6**	-26		Sec 8.4, 5/11
2	a. 19 b. $\dfrac{2}{11}$ c. -2.9		a – b. Sec 8.4, 1/11, 3/11 c. not a video p. 585 #24	**7**	9	Ex 8b, p. 582	
3	-12	Ex 5a, p. 581		**8**	$\dfrac{7}{5}$		Sec 8.4, 7/11
4	-13		Sec 8.4, 4/11	**9**	$90°, 52°$	Ex 11, p. 584	
5	-6	Ex 6a, p. 581		**10**	$30°$		Sec 8.4, 11/11

☐ **Next, insert your homework.** Make sure you attempt all exercises asked of you and show all work, as in the exercises above. Check your answers if possible. Clearly mark any exercises you were unable to correctly complete so that you may ask questions later. DO NOT ERASE YOUR INCORRECT WORK. THIS IS HOW WE UNDERSTAND AND EXPLAIN TO YOU YOUR ERRORS.

Section 8.5 Multiplying and Dividing Real Numbers

Before Class:

☐   Read the objectives on page 592.

☐   Read the **Helpful Hint** boxes on pages 593, 594 and 598.

☐   Read the Multiplying Real Numbers box on page 592 and the Reciprocal box on page 594.

☐   Complete the exercises:

1.   The product of two numbers with the *same* sign is a _____ number. (p.592)

2.   The product of two numbers with *different* signs is a _____ number. (p.592)

3.   What is the base of the exponential expression $(-4)^2$? _____ (p.594)

4.   What is the base of the exponential expression -4^2? _____ (p.594)

5.   What is the reciprocal of $\dfrac{3}{4}$? _____ (p.594)

During Class:

☐   **Write your class notes.**   Neatly write down **all** examples shown as well as key terms or phrases with definitions. If not applicable or if you were absent, watch the Lecture Series (DVD) for this section and do the same (write down the examples shown as well as key terms or phrases). Insert more paper as needed.

Class Notes/Examples	Your Notes

Answers: **1)** positive   **2)** negative   **3)** -4 **4)** 4   **5)** $\dfrac{4}{3}$

Section 8.5 Multiplying and Dividing Real Numbers

Class Notes (continued)	Your Notes

(Insert additional paper as needed.)

Section 8.5 Multiplying and Dividing Real Numbers

Practice:

☐   Complete the Vocabulary and Readiness Check on page 585.

☐   Complete any incomplete exercises below. Check and correct your work using the
     answers and references at the end of this section.

Review this example:	**Your turn:**
1. Multiply. a. $-2\cdot(-14)$ b. $-\dfrac{2}{3}\cdot\dfrac{4}{7}$	**2.** Multiply.
a. $-2\cdot(-14)$	a. $-6(4)$
The product of two numbers with the **same sign** is a positive number.	
$-2\cdot(-14) = \boxed{28}$	b. $-5(-10)$
b. $-\dfrac{2}{3}\cdot\dfrac{4}{7}$	
The product of two numbers with **different signs** is a negative number. Multiply the numerators. Multiply the denominators.	c. $-\dfrac{1}{8}\left(-\dfrac{1}{3}\right)$
$-\dfrac{2}{3}\cdot\dfrac{4}{7} = -\dfrac{2\cdot 4}{3\cdot 7} = \boxed{-\dfrac{8}{21}}$	

Review this example:	**Your turn:**
3. Multiply. a. $(-3)^2$ b. -3^2	**4.** Multiply. a. $(-2)^4$ b. -2^4
a. $(-3)^2 = (-3)(-3) = \boxed{9}$ -3 is the base	
b. $-3^2 = -(3\cdot 3) = \boxed{-9}$ 3 is the base	

Complete this example:	**Your turn:**
5. Find the reciprocal of $-\dfrac{9}{13}$.	**6.** Find the reciprocal of -14.
The reciprocal of $-\dfrac{9}{13}$ is $\boxed{-\dfrac{13}{9}}$ since	
$-\dfrac{9}{13}\cdot -\dfrac{13}{9} = \boxed{}.$	

Section 8.5 Multiplying and Dividing Real Numbers

Review this example:

7. Divide. a. $\dfrac{-100}{5}$ b. $-\dfrac{1}{6} \div \left(-\dfrac{2}{3}\right)$

a. $\dfrac{-100}{5} = \boxed{-20}$ The quotient of two numbers with **different signs** is a negative number.

b. $-\dfrac{1}{6} \div \left(-\dfrac{2}{3}\right) = -\dfrac{1}{6} \cdot \left(-\dfrac{3}{2}\right)$ Multiply the 1st fraction by the reciprocal of the 2nd fraction.

$= \dfrac{3}{12} = \dfrac{\cancel{3}^{1}}{4 \cdot \cancel{3}_{1}} = \boxed{\dfrac{1}{4}}$

The quotient of two numbers with **same signs** is a positive number.

Your turn:

8. Divide.

a. $\dfrac{18}{-2}$

b. $-\dfrac{5}{9} \div \left(-\dfrac{3}{4}\right)$

	Answer	Text Ref	Video Ref		Answer	Text Ref	Video Ref
1	a. 28 b. −20 c. −$\dfrac{8}{21}$	Ex 2, 3 and 4, p. 592		**5**	−$\dfrac{13}{9}$, one	Ex 9d, p. 594	
2	a. −24 b. 50 c. $\dfrac{1}{24}$		Sec 8.5, 2/19, 4/19, 5/19	**6**	−$\dfrac{1}{14}$		Sec 8.5, 10/19
3	a. 9 b. −9	Ex 8c-d, p. 593		**7**	a. −20 b. $\dfrac{1}{4}$	Ex 11b, p. 595, Ex 13, p. 596	
4	a. 16 b. −16		Sec 8.5, 6/19, 7/19	**8**	a. −9 b. $\dfrac{20}{27}$		Sec 8.5, 11/19, 13/19

☐ **Next, insert your homework.** Make sure you attempt all exercises asked of you and show all work, as in the exercises above. Check your answers if possible. Clearly mark any exercises you were unable to correctly complete so that you may ask questions later. DO NOT ERASE YOUR INCORRECT WORK. THIS IS HOW WE UNDERSTAND AND EXPLAIN TO YOU YOUR ERRORS.

Before Class:

☐   Read the objectives on page 605.

☐   Read the **Helpful Hint** boxes on pages 605, 606 and 608.

☐   Read the Commutative Properties box on page 605, the Associative Properties box on page 606, and the Distributive Property of Multiplication Over Addition box on page 607.

☐   Complete the exercises:

_____ 1.   True or False: $6 + (-2) = (-2) + 6$

_____ 2.   True or False: $6 \cdot (-2) = (-2) \cdot 6$

_____ 3.   True or False: $6 - (-2) = (-2) - 6$

_____ 4.   True or False: $6 \div (-2) = (-2) \div 6$

During Class:

☐   **Write your class notes.**   Neatly write down **all** examples shown as well as key terms or phrases with definitions. If not applicable or if you were absent, watch the Lecture Series (DVD) for this section and do the same (write down the examples shown as well as key terms or phrases). Insert more paper as needed.

Class Notes/Examples	**Your Notes**

Answers: **1)** $4 = 4$ so true   **2)** $-12 = -12$ so true   **3)** $8 \neq -8$ **4)** $-3 \neq -\dfrac{1}{3}$

Section 8.6 Properties of Real Numbers

Class Notes (continued)	**Your Notes**

(Insert additional paper as needed.)

Section 8.6 Properties of Real Numbers

Practice:

☐ Complete the Vocabulary and Readiness Check on page 610.

☐ Next, complete any incomplete exercises below. Check and correct your work using the answers and references at the end of this section.

Review this example:	**Your turn:**
1. Use a commutative property to complete each statement: a. $x+5$ b. $3 \cdot x$	2. Use a commutative property to complete each statement.
Change the order.	a. $x+16$ b. xy
a. $x+5 = \boxed{5+x}$	
b. $3 \cdot x = \boxed{x \cdot 3}$	
Review this example:	**Your turn:**
3. Use an associative property to complete the statement: $(-1 \cdot 2) \cdot 5$	4. Use an associative property to complete each statement.
Change the grouping.	a. $(a+b)+c = $ _____
$(-1 \cdot 2) \cdot 5 = \boxed{-1 \cdot (2 \cdot 5)}$	b. $(xy) \cdot z = $ _____
Review this example:	**Your turn:**
5. Simplify the expression: $10+(x+12)$.	6. Use the associative property to simplify the expression $8+(9+b)$.
First, use the commutative property to change the order of x and 12. Second, use the associative property to group the numbers together. Then, simplify.	

$$\overset{\text{Change order}}{} \quad \overset{\text{Change grouping}}{}$$
$$10+(x+12) = 10+(12+x) = (10+12)+x$$

$$10+(x+12) = \boxed{22+x}$$

Section 8.6 Properties of Real Numbers

Complete this example:

7. Use the distributive property to write the expression without parentheses. Then, simplify the result if possible.

   a. $-5(-3+2z)$ b. $-(3+x-w)$

a. $-5(-3+2z)=-5(-3)+-5(2z)=\boxed{15-10z}$

b. $-(3+x-w)=-1(3+x-w)$
            $=(-1)(3)+(-1)(x)-(-1)(w)$

            $=$ ⬭

Your turn:

8. Use the distributive property to write each expression without parentheses. Then, simplify the result if possible.

   a. $-9(4x+8)+2$

   b. $-(r-3-7p)$

Complete this example:

9. Name the property illustrated by the true statement: $-2+2=0$

The (additive inverse property) is used when you add _____.

Your turn:

10. Name the property illustrated by the true statement: $6\cdot\dfrac{1}{6}=1$

	Answer	Text Ref	Video Ref		Answer	Text Ref	Video Ref
1	a. $5+x$ b. $x\cdot 3$	Ex 1, p. 605		6	$17+b$		Sec 8.6, 6/15
2	a. $16+x$ b. yx		Sec 8.6, 1 – 2/15	7	a. $15-10z$ b. $-3-x+z$	Ex 8,11, p. 607,608	
3	$-1\cdot(2\cdot 5)$	Ex 2b, p. 606		8	a. $-36x-70$ b. $-r+3+7p$		Sec 8.6, 9 – 10/15
4	a. $a+(b+c)$ b. $x\cdot(yz)$		Sec 8.6, 4 – 5/15	9	Additive inverse, opposites	Ex 20, p. 609	
5	$22+x$	Ex 5, p. 607		10	Multiplicative inverse		Sec 8.6, 13/15

☐   **Next, insert your homework.** Make sure you attempt all exercises asked of you and show all work, as in the exercises above. Check your answers if possible. Clearly mark any exercises you were unable to correctly complete so that you may ask questions later. DO NOT ERASE YOUR INCORRECT WORK. THIS IS HOW WE UNDERSTAND AND EXPLAIN TO YOU YOUR ERRORS.

Section 8.7 Simplifying Expressions

Before Class:

☐   Read the objectives on page 614.

☐   Read the **Helpful Hint** boxes on page 614, 615, and 617.

☐   Read the Like Terms/Unlike Terms chart on page 615.

☐   Complete the exercises:

1.   Identify the numerical coefficient of each term. (p. 614)

   a. x b. $-x$ c. $2x$ d. $-y^2$

2.   Are $3x$ and $-4x$ like terms? (p. 615)   3.   Are $3x$ and $-4x^2$ like terms?

4.   Are $3xy$ and $-4yx$ like terms?   5.   Are $3x^2y$ and $-4xy^2$ like terms?

During Class:

☐   **Write your class notes.**   Neatly write down **all** examples shown as well as key terms or phrases with definitions. If not applicable or if you were absent, watch the Lecture Series (DVD) for this section and do the same (write down the examples shown as well as key terms or phrases). Insert more paper as needed.

Class Notes/Examples	Your Notes

Answers: **1)**   a. 1   b. -1 c. 2   d. -1 **2)** yes   **3)** no   **4)** yes   **5)** no

Section 8.7 Simplifying Expressions

Class Notes (continued)	Your Notes

(Insert additional paper as needed.)

Section 8.7 Simplifying Expressions

Practice:

☐   Complete the Vocabulary and Readiness Check on page 619.

☐   Next, complete any incomplete exercises below. Check and correct your work using the answers and references at the end of this section.

Review this example:
1. Simplify each expression by combining like terms.   a. $7x - 3x$ b. $10y^2 + y^2$

a. $7x - 3x = (7 - 3)x$

$= \boxed{4x}$ Use the distributive property.

b. $10y^2 + y^2 = (10 + 1)y^2$

$= \boxed{11y^2}$

Your turn:
2. Simplify each expression by combining any like terms.

a. $3x + 2x$

b. $m - 4m + 2m - 6$

Complete this example:
3. Simplify the expression by combining like terms: $2x + 3x + 5 + 2$

Add numerical coefficients of like terms.

$2x + 3x + 5 + 2 = (2 + 3)x + (5 + 2)$

$= \bigcirc$

Your turn:
4. Simplify each expression by combining any like terms.

a. $8x^3 + x^3 - 11x^3$

b. $6x + 0.5 - 4.3x - 0.4x + 3$

Review this example:
5. Subtract $4x - 2$ from $2x - 3$.

Translate: $(2x - 3) - (4x - 2)$

$(2x - 3) - (4x - 2) = 2x - 3 - 4x + 2$ Distribute.

$= \boxed{-2x - 1}$ Simplify.

Your turn:
6. Subtract $5m - 6$ from $m - 9$.

217

Section 8.7 Simplifying Expressions

Complete this example:

7. Use the distributive property to simplify. Then, combine like terms.

$$3(2x-5)+1$$

$$3(2x - 5) + 1 = 6x - 15 + 1$$

$$= \bigcirc$$

Your turn:

8. Remove parentheses and simplify the expression.

$$5(x+2)-(3x-4)$$

Review this example:

9. Write as an algebraic expression and simplify: *Four times the sum of a number and 3.*

Four times the sum of a number and 3

$$\begin{array}{ccc} \downarrow & \downarrow & \\ 4 & \cdot & (x + 3) \end{array}$$

Now, use the distributive property to simplify.

$$4 \cdot (x + 3) = 4(x + 3)$$

$$= \boxed{4x+12}$$

Your turn:

10. Write as an algebraic expression.
    Twice a number, decreased by four.

	Answer	Text Ref	Video Ref		Answer	Text Ref	Video Ref
1	a. $4x$ b. $11y^2$	Ex 3ab, p. 615		6	$-4m-3$		Sec 8.7, 12/14
2	a. $5x$ b. $-m-6$		a. Sec 8.7, 8/14 b. not a video p. 619 #7	7	$6x-14$	Ex 11, p. 617	
3	$5x+7$	Ex 4, p. 616		8	$2x+14$		Sec 8.7, 11/14
4	a. $-2x^3$ b. $1.3x+3.5$		Sec 8.7, 9 – 10/14	9	$4x+12$	Ex 19, p. 618	
5	$-2x-1$	Ex 15, p. 617		10	$2x-4$		Sec 8.7, 13/14

☐ **Next, insert your homework.** Make sure you attempt all exercises asked of you and show all work, as in the exercises above. Check your answers if possible. Clearly mark any exercises you were unable to correctly complete so that you may ask questions later. DO NOT ERASE YOUR INCORRECT WORK. THIS IS HOW WE UNDERSTAND AND EXPLAIN TO YOU YOUR ERRORS.

Preparing for the Chapter 8 Test

Start preparing for your Chapter 8 Test as soon as possible. Pay careful attention to any instructor discussion about this test, especially discussion on what sections you will be responsible for, etc.

☐ Work the Chapter 8 Vocabulary Check on page 624.

☐ Read both columns (Definitions and Concepts, and Examples) of the Chapter 8 Highlights starting on page 624.

☐ Read your Class Notes/Examples for each section covered on your Chapter 8 Test. Look for any unresolved questions you may have.

☐ Complete as many of the Chapter 8 Review exercises as possible (pages 628 – 632). Remember, the odd answers are in the back of your text.

☐ **Most important:** Place yourself in "test" conditions (see below) and work the Chapter 8 Test (pages 633 – 634) as a practice test the day before your actual test. To honestly assess how you are doing, try the following:
- Work on a few blank sheets of paper.
- Give yourself the same amount of time you will be given for your actual test.
- Complete this Chapter 8 Practice Test without using your notes or your text.
- If you have any time left after completing this practice test, check your work and try to find any errors on your own.
- Once done, use the back of your book to check ALL answers.
- Try to correct any errors on your own.
- Use the Chapter Test Prep Video (CTPV) to correct any errors you were unable to correct on your own. You can find these videos in the Interactive DVD Lecture Series, in MyMathLab, and on YouTube. Search MartinGayBasicMath and click "Channels."

I wish you the best of luck….Elayn Martin-Gay

Section 9.1 The Addition Property of Equality

Before Class:

☐   Read the objectives on page 638.

☐   Read the **Helpful Hint** boxes on pages 638 and 640.

☐   Complete the exercises:

     Simplify each expression.

     1. $5x - x + 3x$ 2. $\dfrac{1}{2}x + 3 + \dfrac{5}{2}x - 6$ 3. $2(x - 5)$

During Class:

☐   **Write your class notes.** Neatly write down **all** examples shown as well as key terms or phrases with definitions. If not applicable or if you were absent, watch the Lecture Series (DVD) for this section and do the same (write down the examples shown as well as key terms or phrases). Insert more paper as needed.

Class Notes/Examples	**Your Notes**

Answers: **1)** $7x$ **2)** $3x - 3$ **3)** $2x - 10$

Section 9.1 The Addition Property of Equality

Class Notes (continued)	**Your Notes**

(Insert additional paper as needed.)

Section 9.1 The Addition Property of Equality

Practice:

☐   Complete the Vocabulary and Readiness Check on page 643.

☐   Next, complete any incomplete exercises below. Check and correct your work using the answers and references at the end of this section.

Review this example:

1. Solve $x - 7 = 10$. Check the solution.

Rewrite the equation in the form x = number.

$$x - 7 = 10$$
$$x - 7 + 7 = 10 + 7 \quad \text{Add } 7 \text{ to both sides.}$$
$$\boxed{x = 17} \quad \text{Simplify.}$$

Check:

$$x - 7 = 10$$
$$17 - 7 \overset{?}{=} 10 \quad \text{Replace } x \text{ with } 17.$$
$$10 = 10 \quad \text{This is a true statement.}$$

The solution is 17.

Your turn:

2. Solve $x - 2 = -4$. Check the solution.

Review this example:

3. Solve $5t - 5 = 6t$. Check the solution.

We want all terms containing t on one side of the equation and numbers on the other side.

$$5t - 5 = 6t$$
$$5t - 5 - 5t = 6t - 5t \quad \text{Subtract } 5t \text{ from both sides.}$$
$$\boxed{-5 = t} \quad \text{Combine like terms.}$$

Check:

$$5t - 5 = 6t$$
$$5(-5) - 5 \overset{?}{=} 6(-5) \quad \text{Replace } t \text{ with } -5.$$
$$-25 - 5 \overset{?}{=} -30$$
$$-30 = -30 \quad \text{This is a true statement.}$$

The solution is -5.

Your turn:

4. Solve $5b - 0.7 = 6b$. Check the solution.

Section 9.1 The Addition Property of Equality

Complete this example:

5. Solve: $2x + 3x - 5 + 7 = 10x + 3 - 6x - 4$

First, simplify both sides of the equation.

$5x + 2 = 4x - 1$ Combine like terms.

   Subtract $4x$ from both sides.

$5x + 2 - 4x = 4x - 1 - 4x$

$x + 2 = -1$ Combine like terms.

$x + 2 - 2 = -1 - 2$ Subtract 2 from both sides.

$\boxed{x = -3}$ Combine like terms.

Check:

Your turn:

6. Solve: $13x - 9 + 2x - 5 = 12x - 1 + 2x$

Check:

Complete this example:

7. Solve: $6(2a - 1) - (11a + 6) = 7$ Check.

First, apply the distributive property.

$6(2a) + 6(-1) - 1(11a) - 1(6) = 7$

$12a - 6 - 11a - 6 = 7$ Multiply.

$a - 12 = 7$ Combine like terms.

$a - 12 + 12 = 7 + 12$ Add 12 to both sides.

$\boxed{a = 19}$ Simplify.

Check:

Your turn:

8. Solve: $15 - (6 - 7k) = 2 + 6k$ Check.

Check:

	Answer	Text Ref	Video Ref		Answer	Text Ref	Video Ref
1	17	Ex 1, p. 639		5	−3	Ex 5, p. 641	
2	−2		Sec 9.1, 1/8	6	13		Sec 9.1, 5/8
3	−5	Ex 4, p. 640		7	19	Ex 6, p. 641	
4	−0.7		Sec 9.1, 3/8	8	−7		Sec 9.1, 6/8

☐   **Next, insert your homework.** Make sure you attempt all exercises asked of you and show all work, as in the exercises above. Check your answers if possible. Clearly mark any exercises you were unable to correctly complete so that you may ask questions later. DO NOT ERASE YOUR INCORRECT WORK. THIS IS HOW WE UNDERSTAND AND EXPLAIN TO YOU YOUR ERRORS.

Section 9.2 The Multiplication Property of Equality

Before Class:

☐   Read the objectives on page 647.

☐   Read the **Helpful Hint** box on page 652.

☐   Read the **Multiplication Property of Equality** box on page 647.

☐   Complete the exercises:

Perform the indicated operation.

1. $\dfrac{-7}{-7} =$ _____

2. $\dfrac{4.2}{4.2} =$ _____

3. $\dfrac{1}{3} \cdot 3 =$ _____

4. $\dfrac{1}{5} \cdot 5 =$ _____

5. $-\dfrac{2}{3} \cdot -\dfrac{3}{2} =$ _____

6. $-\dfrac{7}{2} \cdot -\dfrac{2}{7} =$ _____

During Class:

☐   **Write your class notes.** Neatly write down **all** examples shown as well as key terms or phrases with definitions. If not applicable or if you were absent, watch the Lecture Series (DVD) for this section and do the same (write down the examples shown as well as key terms or phrases). Insert more paper as needed.

Class Notes/Examples	**Your Notes**

Answers: **1-6)** all answers are 1

Section 9.2 The Multiplication Property of Equality

Class Notes (continued)	**Your Notes**

(Insert additional paper as needed.)

Section 9.2 The Multiplication Property of Equality

Practice:

☐   Complete the Vocabulary and Readiness Check on page 653.

☐   Next, complete any incomplete exercises below. Check and correct your work using the answers and references at the end of this section.

Complete this example:	Your turn:
1. Solve: $\dfrac{5}{2}x = 15$	**2.** Solve: $\dfrac{2}{3}x = -8$

$$\frac{5}{2}x = 15$$

$\dfrac{2}{5} \cdot \left(\dfrac{5}{2}x\right) = \dfrac{2}{5} \cdot 15$ Multiply both sides by the reciprocal of $\dfrac{5}{2}$, which is $\dfrac{2}{5}$.

$\left(\dfrac{2}{5} \cdot \dfrac{5}{2}\right)x = \dfrac{2}{5} \cdot 15$ Apply the associative property.

$1x = 6$

$\boxed{x = 6}$ Simplify.

Check:

Check:

Complete this example:	Your turn:
3. Solve: $7x - 3 = 5x + 9$	**4.** Solve: $8x + 20 = 6x + 18$

$7x - 3 = 5x + 9$

$7x - 3 - 5x = 5x + 9 - 5x$ Subtract $5x$ both sides.

$2x - 3 = 9$ Simplify.

$2x - 3 + 3 = 9 + 3$ Add 3 to both sides.

$2x = 12$ Simplify.

$\dfrac{2x}{2} = \dfrac{12}{2}$ Divide both sides by 2.

$\boxed{x = 6}$ Simplify.

Check:

Check:

Section 9.2 The Multiplication Property of Equality

Review this example:

5. Solve: $5(2x+3) = -1+7$

$5(2x+3) = -1+7$	Apply the distributive
$5(2x)+5(3) = -1+7$	property.
$10x+15 = 6$	Multiply. Simplify.
$10x+15-15 = 6-15$	Subtract 15 from both sides.
$10x = -9$	Simplify.
$\dfrac{10x}{10} = -\dfrac{9}{10}$	Divide both sides by 10.
$x = -\dfrac{9}{10}$	Simplify.

Your turn:

6. Solve: $9(3x+1) = 4x-5x$

Review this example:

7. If x is the first of three consecutive integers, express the sum of the three integers in terms of x. Simplify if possible.

UNDERSTAND.

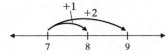

Let x = the unknown number.

TRANSLATE.

first integer	+	second integer	+	third integer

$$x \quad + \quad (x+1) \quad + \quad (x+2)$$

Simplies to $3x+3$.

Your turn:

8. If x is the first of four consecutive integers, express the sum of the first integer and the third integer as an algebraic expression containing the variable x.

	Answer	Text Ref	Video Ref		Answer	Text Ref	Video Ref
1	6	Ex 1, p. 647		5	$-\dfrac{9}{10}$	Ex 10, p.651	
2	−12		Sec 9.2, 2/8	6	$-\dfrac{9}{28}$		Sec 9.2, 7/8
3	6	Ex 9, p.650		7	$3x+3$	Ex 11, p.651	
4	−1		Sec 9.2, 5/8	8	$2x+2$		Sec 9.2, 8/8

☐   **Next, insert your homework.** Make sure you attempt all exercises asked of you and show all work, as in the exercises above. Check your answers if possible. Clearly mark any exercises you were unable to correctly complete so that you may ask questions later. DO NOT ERASE YOUR INCORRECT WORK. THIS IS HOW WE UNDERSTAND AND EXPLAIN TO YOU YOUR ERRORS.

Section 9.3 Further Solving Linear Equations

Before Class:

☐   Read the objectives on page 656.

☐   Read the **Helpful Hint** boxes on pages 657, 658, 659.

☐   Complete the exercises:

Fill in the table with the opposite (additive inverse), the reciprocal (multiplicative inverse), or the expression. Assume that the value of each expression is not 0.

	1.	2.	3.	4.
Expression	9	$\dfrac{3}{4}$	$2x$	
Opposite				$-3y$
Reciprocal				

During Class:

☐   **Write your class notes.** Neatly write down **all** examples shown as well as key terms or phrases with definitions. If not applicable or if you were absent, watch the Lecture Series (DVD) for this section and do the same (write down the examples shown as well as key terms or phrases). Insert more paper as needed.

Class Notes/Examples	**Your Notes**

Answers: **1)** $-9; \dfrac{1}{9}$ **2)** $-\dfrac{3}{4}; \dfrac{4}{3}$ **3)** $-2x; \dfrac{1}{2x}$ **4)** $3y; \dfrac{1}{3y}$

Section 9.3 Further Solving Linear Equations

Class Notes (continued)

Your Notes

(Insert additional paper as needed.)

Section 9.3 Further Solving Linear Equations

Practice:

☐   Complete the Vocabulary and Readiness Check on page 661.

☐   Next, complete any incomplete exercises below. Check and correct your work using the answers and references at the end of this section.

Review this example:

1. Solve: $4(2x-3)+7=3x+5$

$8x-12+7=3x+5$ Distribute.

$8x-5=3x+5$ Combine like terms.

$8x-5-3x=3x+5-3x$ Subtract $3x$ from both

$5x-5=5$ sides, then simplify.

$5x-5+5=5+5$ Add 5 to both sides.

$5x=10$ Simplify.

$\dfrac{5x}{5}=\dfrac{10}{5}$ Divide both sides by 5.

$\boxed{x=2}$ Simplify.

Check:

$4[2(2)-3]+7=3(2)+5$

$4(4-3)+7=6+5$

$4(1)+7=11$

$11=11$

Your turn:

2. Solve: $5(2x-1)-2(3x)=1$

Complete this example:

3. Solve: $\dfrac{x}{2}-1=\dfrac{2}{3}x-3$

First we clear the fractions by multiplying both sides of the equation by the LCD of 6.

$6\left(\dfrac{x}{2}-1\right)=6\left(\dfrac{2}{3}x-3\right)$

$6\left(\dfrac{x}{2}\right)-6(1)=6\left(\dfrac{2}{3}x\right)-6(3)$ Distribute.

$3x-6=4x-18$ Simplify.

$3x-6-3x=4x-18-3x$ Subtract $3x$ from

$\underline{\quad}=\underline{\quad}-18$ both sides, simplify.

$\underline{\quad}+18=\underline{\quad}-18+18$ Add 18 to both

$\boxed{\underline{\quad}=x}$ sides, simplify.

Your turn:

4. Solve: $\dfrac{x}{2}-1=\dfrac{x}{5}+2$

231

Section 9.3 Further Solving Linear Equations

Review this example:

5. Solve: $0.25x + 0.10(x - 3) = 1.1$

First, we clear the equation of decimals by multiplying both sides of the equation by 100.

Move each decimal 2 places to the right.

$25x + 10(x - 3) = 110$

$25x + 10x - 30 = 110$ Distribute.

$35x + 30 = 110$ Combine like terms.

$35x - 30 + 30 = 110 + 30$ Add 30 to both sides,

$35x = 140$ then simplify.

$\dfrac{35x}{35} = \dfrac{140}{35}$ Divide both sides by 35, then simplify.

$x = 4$

Your turn:

6. Solve: $0.50x + 0.15(70) = 35.5$

Review this example:

7. Solve: $3(x - 4) = 3x - 12$

$3x - 12 = 3x - 12$ Distribute.

$3x - 12 - 3x = 3x - 12 - 3x$ Subtract 3x from both

$-12 = 12$ sides, then simplify.

The left side is identical to the right side.

Every real number is a solution.

Your turn:

8. Solve: $2(x + 3) - 5 = 5x - 3(1 + x)$

	Answer	Text Ref	Video Ref		Answer	Text Ref	Video Ref
1	$x = 2$	Ex 1, p. 656		5	$x = 4$	Ex 5, p. 659	
2	$x = \dfrac{3}{2}$		Sec 9.3, 1/5	6	$x = 50$		Sec 9.3, 3/5
3	$x = 12$	Ex 3, p. 657–658		7	all real numbers	Ex 7, p. 660	
4	$x = 10$		Sec 9.3, 2/5	8	no solution		Sec 9.3, 5/5

☐   **Next, insert your homework.** Make sure you attempt all exercises asked of you and show all work, as in the exercises above. Check your answers if possible. Clearly mark any exercises you were unable to correctly complete so that you may ask questions later. DO NOT ERASE YOUR INCORRECT WORK. THIS IS HOW WE UNDERSTAND AND EXPLAIN TO YOU YOUR ERRORS.

Section 9.4 A Further Introduction to Problem Solving

Before Class:

☐   Read the objectives on page 667.

☐   Read the **Helpful Hint** boxes on page 667, 668, 669, and 673.

☐   Complete the exercises:

1.   Translate using x for the number: Twice the difference of a number and 8 is equal to three times the number.

2.   Translate using x for the number: A number added to 50 more than the number is 200.

3.   Let x be an even integer. What are the next two consecutive even integers?

4.   In an isosceles triangle the two base angles have the _____degree measure.

5.   Recall the sum of the measure of the angles of a triangle equals _____.

During Class:

☐   **Write your class notes.** Neatly write down **all** examples shown as well as key terms or phrases with definitions. If not applicable or if you were absent, watch the Lecture Series (DVD) for this section and do the same (write down the examples shown as well as key terms or phrases). Insert more paper as needed.

Class Notes/Examples	**Your Notes**

Answers: **1)** $2(x-8)=3x$ **2)** $x+(x+50)=200$ **3)** $x+2, x+4$ **4)** same   **5)** 180

Section 9.4 A Further Introduction to Problem Solving

Class Notes (continued)	Your Notes

(Insert additional paper as needed.)

Section 9.4 A Further Introduction to Problem Solving

Practice:

☐   Complete the Vocabulary and Readiness Check on page 674.

☐   Next, complete any incomplete exercises below. Check and correct your work using the answers and references at the end of this section.

Review this example:	**Your turn:**

1. Twice the sum of a number and 4 is the same as four times the number decreased by 12. Find the number.

2. Twice the difference of a number and 8 is equal to three times the sum of the number and 3. Find the number.

UNDERSTAND. Let x = the unknown number.

TRANSLATE. $2(x+4) = 4x - 12$

SOLVE and INTERPRET.

$2(x+4) = 4x - 12$

$2x + 8 = 4x - 12$ Distribute.

$2x + 8 - 4x = 4x - 12 - 4x$ Subtract $4x$ from both sides.

$-2x + 8 = -12$

$-2x + 8 - 8 = -12 - 8$ Then subtract 8 from both sides.

$-2x = -20$

$\dfrac{-2x}{-2} = \dfrac{-20}{-2}$ Divide both sides by –2.

$x = 10$

The number is 10.

Check:
$2(10 + 4) = 4(10) - 12$
$2(14) = 40 - 12$
$28 = 28$

Complete this Example:	**Your turn:**

3. The 111th Congress had a total of 434 Democrats and Republicans. There were 78 more Democrats than Republicans. Find the number of representatives from each party.

4. The area of the Sahara Desert is 7 times the area of the Gobi Desert. If the sum of their areas is 4,000,000 square miles, find the area of each desert.

UNDERSTAND: x = number of Republicans
$x + 78$ = number of Democrats

TRANSLATE: $x + (x + 78) = 434$

SOLVE and INTERPRET:

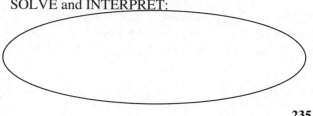

Section 9.4 A Further Introduction to Problem Solving

Complete this example:

5. If the two walls of the Vietnam Veterans Memorial in Washington, D.C., were connected, an isosceles triangle would be formed. The measure of the third angle is 97.5° more than the measure of either of the two equal angles. Find the measure of the third angle.

UNDERSTAND.

x = degree measure of one base angle
x = degree measure of the second base angle
$x + 97.5$ = degree measure of the third angle

TRANSLATE and SOLVE.

$x + x + (x + 97.5) = 180$ Combine like terms.
$3x + 97.5 = 180$
$3x + 97.5 - 97.5 = 180 - 97.5$ Subtract 97.5 from both sides.
$3x = 82.5$

$x =$

INTERPRET your answer here:

Your turn:

6. The measures of the angles of a triangle are 3 consecutive even integers. Find the measure of each angle.

	Answer	Text Ref	Video Ref		Answer	Text Ref	Video Ref
1	$x = 10$	Ex 2, p. 668		4	Sahara: 3,500,000 sq mi Gobi: 500,000 sq mi		Sec 9.4, 2/4
2	$x = -25$		Sec 9.4, 1/4	5	equal angles: 27.5°, 27.5° 3rd angle: 125° $27.5 + 27.5 + 125 = 180$	Ex 6, p. 671	
3	178 Republicans and 256 Democrats $178 + 256 = 434$ total	Ex 4, p. 669		6	Let x = the 1st even integer. $x = 58$ $58°, 60°, 62°$		Sec 9.4, 4/4

☐ **Next, insert your homework.** Make sure you attempt all exercises asked of you and show all work, as in the exercises above. Check your answers if possible. Clearly mark any exercises you were unable to correctly complete so that you may ask questions later. DO NOT ERASE YOUR INCORRECT WORK. THIS IS HOW WE UNDERSTAND AND EXPLAIN TO YOU YOUR ERRORS.

Section 9.5 Formulas and Problem Solving

Before Class:

☐   Read the objectives on page 683.

☐   Read the **Helpful Hint** boxes on pages 684 and 688.

☐   Complete the exercises:

Evaluate each expression for the given values.

1. $2W + 2L$; $W = 6$ and $L = 10$ 2. $\frac{1}{2}Bh$; $B = 8$ and $h = 24$

3. πr^2; $r = 4$ 4. $r \cdot t$; $r = 18$ and $t = 2$

During Class:

☐   **Write your class notes.** Neatly write down **all** examples shown as well as key terms or phrases with definitions. If not applicable or if you were absent, watch the Lecture Series (DVD) for this section and do the same (write down the examples shown as well as key terms or phrases). Insert more paper as needed.

Class Notes/Examples	**Your Notes**

Answers: **1)** 32   **2)** 96   **3)** 16π **4)** 36

Section 9.5 Formulas and Problem Solving

Class Notes (continued)	Your Notes

(Insert additional paper as needed.)

Practice:

☐   Complete any incomplete exercises below. Check and correct your work using the answers and references at the end of this section.

Review this example:	**Your turn:**
1. The average maximum temperature for January in Algiers, Algeria, is 59° Fahrenheit. Find the equivalent temperature in degrees Celsius.	2. Convert Nome, Alaska's 14° F high temperature to Celsius.

UNDERSTAND.

Let C = temperature in degrees Celsius.
Let F = temperature in degrees Fahrenheit.

TRANSLATE. Use $F = \dfrac{9}{5}C + 32$; $F = 59$.

SOLVE. Let F = 59.

$$59 = \frac{9}{5}C + 32$$

$59 - 32 = \dfrac{9}{5}C + 32 - 32$ Subtract 32 from both sides, combine like terms.

$$27 = \frac{9}{5}C$$

$\dfrac{5}{9} \cdot 27 = \dfrac{5}{9} \cdot \dfrac{9}{5}C$ Multiply both sides by $\dfrac{5}{9}$, then simplify.

$\boxed{15 = C}$

Review this Example:	**Your turn:**
3. Solve $y = mx + b$ for x.	4. Solve $V = lwh$ for w.

$y = mx + b$

$y - b = mx + b - b$ Subtract b from both sides.

$y - b = mx$ Combine like terms.

$\dfrac{y - b}{m} = \dfrac{mx}{m}$ Divide both sides by m.

$\boxed{\dfrac{y - b}{m} = x}$ Simplify.

Section 9.5 Formulas and Problem Solving

Complete this example:

5. The length of a rectangular road sign is 2 feet less than three times its width. Find the dimensions if the perimeter is 28 feet.

UNDERSTAND. Use the formula $P = 2l + 2w$. Let $w =$ the width of the rectangular sign, and $3w - 2 =$ the length of the sign.

TRANSLATE. $l = 3w - 2$ and $P = 28$
$P = 2l + 2w$
$28 = 2(3w - 2) + 2w$

SOLVE and INTERPRET.

$28 = 6w - 4 + 2w$	Distribute.
$28 = 8w - 4$	Combine like terms.
$28 + 4 = 8w - 4 + 4$	Add 4 to both sides.
$32 = 8w$	Simplify.
$\dfrac{32}{8} = \dfrac{8w}{8}$	Divide both sides by 8.
$\underline{\quad} = w$	

The width is _____ feet.

The length is _____ feet.

Check:
$P = 2l + 2w$

$P = 2(\underline{\quad}) + 2(\underline{\quad})$

$P = \underline{\quad}$ feet

Your turn:

6. An architect designs a rectangular flower garden such that the width is exactly two-thirds of the length. If 260 feet of antique picket fencing are to be used to enclose the garden, find the dimensions of the garden.

	Answer	Text Ref	Video Ref		Answer	Text Ref	Video Ref
1	$C = 15$	Ex 3, p. 685		4	$w = \dfrac{V}{lh}$		Sec 9.5, 4/5
2	$-10°C$		Sec 9.5, 1/5	5	width: 4 ft; length: 10 ft	Ex 4, p. 686	
3	$\dfrac{y - b}{m} = x$	Ex 6, p. 688		6	length: 78 ft; width: 52 ft		Sec 9.5, 3/5

☐ **Next, insert your homework.** Make sure you attempt all exercises asked of you and show all work, as in the exercises above. Check your answers if possible. Clearly mark any exercises you were unable to correctly complete so that you may ask questions later. DO NOT ERASE YOUR INCORRECT WORK. THIS IS HOW WE UNDERSTAND AND EXPLAIN TO YOU YOUR ERRORS.

Section 9.6 Percent and Mixture Problem Solving

Before Class:

☐   Read the objectives on page 696.

☐   Read the **General Strategy for Problem Solving** box on page 696.

    1.   Write 35% as a decimal. _____

    2.   Write 0.75 as a percent. _____

    3.   Write 1.2 as a percent. _____

    4.   Fill in the blanks with **percent**, **base**, or **amount**.

    In the statement "10% of 80 is 8," the number 8 is called the _____ , 80 is

    called the _____ and 10 is the _____ .

During Class:

☐   **Write your class notes.** Neatly write down **all** examples shown as well as key terms or phrases with definitions. If not applicable or if you were absent, watch the Lecture Series (DVD) for this section and do the same (write down the examples shown as well as key terms or phrases). Insert more paper as needed.

Class Notes/Examples	**Your Notes**

Answers: **1)** 0.35   **2)** 75%   **3)** 120%   **4)** amount, base, and percent

Section 9.6 Percent and Mixture Problem Solving

Class Notes (continued)	**Your Notes**

(Insert additional paper as needed.)

Section 9.6 Percent and Mixture Problem Solving
_____

Practice:

☐   Complete the Vocabulary and Readiness Check on page 702.

☐   Next, complete any incomplete exercises below. Check and correct your work using the
answers and references at the end of this section.

Complete this example:

1. Cells Phones Unlimited recently reduced the
price of a $140 phone by 20%. What is the
discount and the new price?

discount = percent · original price
    ↓   ↓   ↓

$\boxed{\text{discount}}$ =   20% · $140   Write 20%
                                          as a decimal
        = 0.20 · $140   and multiply.

        = $28

new price = original price – discount
    ↓   ↓   ↓

$\boxed{\text{new price}}$ = $140   – $28

The new price is ⬭.

Your turn:

2. A birthday celebration meal is
$40.50 including tax. Find the total
cost if a 15% tip is added to the cost.

Review this example:

3. The fastest growing sector of digital theater
screens is 3D. Find the number of digital 3D
screens in the United States and Canada last
year if after a 134% increase, the number this
year is 3548.

# of screens last year + increase = # of screens now
        ↓   ↓   ↓
        x + $1.34x$ =   3548

$2.34x = 3548$ There were approximately

$x = \dfrac{3548}{2.34}$ _____   digital 3D
                             screens last year.

$x \approx 1516$

Your turn:

4. Find last year's salary if after a 4%
pay raise, this year's salary is
$44,200.

243

Section 9.6 Percent and Mixture Problem Solving

Review this example:

5. A chemist needs 12 liters of a 50% acid s solution. The stockroom has only 40% and 70% solutions. How much of each solution should be mixed together to form 12 liters of a 50% solution?

UNDERSTAND.

x = number of liters of 40% solution

$12 - x$ = number of liters of 70% solution

TRANSLATE .

	No. of Liters	·	Acid Strength	=	Amount of Acid
40% Solution	x		40%		$0.40x$
70% Solution	$12 - x$		70%		$0.70(12 - x)$
50% Solution Needed	12		50%		$0.50(12)$

SOLVE.

$0.40x + 0.70(12 - x) = 0.50(12)$

$0.4x + 8.4 - 0.7x = 6$ Distribute.

_____   Combine like terms.

_____   Subtract ____ from both sides.

_____   Divide both sides by

____.

$\boxed{x =}$

____ liters of 40% solution

____ liters of 70% solution

Your turn:

6. How much of an alloy that is 20% copper should be mixed with 200 ounces of an allow that is 50% copper in order to get an alloy that is 30% copper?

	Answer	Text Ref	Video Ref		Answer	Text Ref	Video Ref
1	*Disct* : $28 *New* : $112	Ex 4, p. 698		4	$42,500		Sec 9.6, 5/6
2	$46.58		Sec 9.6, 3/6	5	8l of 40%; 4l of 70%	Ex 7, p. 700	
3	≈ 1516	Ex 6, p. 700		6	400 oz		Sec 9.6, 6/6

☐   **Next, insert your homework.** Make sure you attempt all exercises asked of you and show all work, as in the exercises above. Check your answers if possible. Clearly mark any exercises you were unable to correctly complete so that you may ask questions later. DO NOT ERASE YOUR INCORRECT WORK. THIS IS HOW WE UNDERSTAND AND EXPLAIN TO YOU YOUR ERRORS.

Section 9.7 Linear Inequalities and Problem Solving

Before Class:

☐   Read the objectives on page 708.

☐   Read the **Helpful Hint** boxes on pages 709 and 711.

☐   Complete the exercises:

1.   Which of the following numbers are solutions to the inequality $x < -2$?

| −4 | −3 | −2 | 0 | 2 | 3 |

2.   Which of the following numbers are solutions to the inequality $x \geq -2$?

| −4 | −3 | −2 | 0 | 2 | 3 |

3.   Which of the following numbers are solutions to the inequality $-4 < x \leq 2$?

| −4 | −3 | −2 | 0 | 2 | 3 |

During Class:

☐   **Write your class notes.** Neatly write down **all** examples shown as well as key terms or phrases with definitions. If not applicable or if you were absent, watch the Lecture Series (DVD) for this section and do the same (write down the examples shown as well as key terms or phrases). Insert more paper as needed.

Class Notes/Examples	Your Notes

Answers: **1)** −4 and −3   **2)** −2, 0, 2, and 3   **3)** −3, −2, 0, and 2

Section 9.7 Linear Inequalities and Problem Solving

Class Notes (continued)	Your Notes

(Insert additional paper as needed.)

Section 9.7 Linear Inequalities and Problem Solving

Practice:

☐   Complete the Vocabulary and Readiness Check on page 714.

☐   Next, complete any incomplete exercises below. Check and correct your work using the answers and references at the end of this section.

Review this example:

1. Graph: $-4 < x \le 2$

This inequality is read as, "–4 is less than x and x is less than or equal to 2."

Place an open circle on –4. The open circle on –4 indicates –4 is not part of the graph. Then, place a closed circle on 2. The closed circle on 2 indicates that 2 is part of the graph. Finally, shade all numbers between –4 and 2.

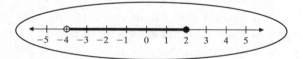

Your turn:

2. Graph the inequality on the number line.

$x \le -1$

Review this example:

3. Solve $x + 4 \le -6$. Graph the solution set. Write the answer using solution set notation.

$$x + 4 \le -6$$
$$x + 4 - 4 \le -6 - 4$$

Subtract 4 from both sides and simplify.

$$x \le -10$$

To graph the solutions of $x \le -10$, place a closed circle on -10. The closed circle indicates that -10 is a solution to the inequality. Then, shade the numbers to the left of -10 to indicate all numbers less than -10.

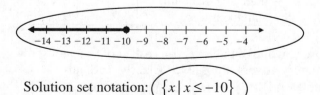

Solution set notation: $\left\{ x \mid x \le -10 \right\}$

Your turn:

4. Solve each inequality. Graph the solution set. Write each answer using solution set notation.

a. $x - 2 \ge -7$

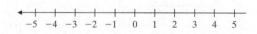

b. $-8x \le 16$

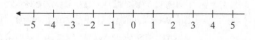

247

Section 9.7 Linear Inequalities and Problem Solving

Complete this example:

5. Solve $-5x+7 < 2(x-3)$. Graph the solution set.

$-5x+7 < 2(x-3)$	Distribute.
$-5x+7 < 2x-6$	Subtract $2x$ from both sides and simplify.
$-5x+7-2x < 2x-6-2x$	
$-7x+7 < -6$	Subtract 7 from both sides and simplify.
$-7x+7-7 < -6-7$	
$-7x < -13$	Complete the last step.

x

Your turn:

6. Solve each inequality. Write each answer using solution set notation.

$$3(x+2)-6 > -2(x-3)+14$$

Complete this example:

7. 12 subtracted from 3 times a number is less than 21. Find all numbers that make this statement true.

Let $x =$ a number, and translate the statement.

$3x-12 < 21$	Add 12 to both sides and simplify.
$3x-12+12 < 21+12$	
$3x < 33$	
x	Complete the last step.

Your turn:

8. Six more than twice a number is greater than negative fourteen. Find all numbers that make this statement true.

	Answer	Text Ref	Video Ref		Answer	Text Ref	Video Ref
1		Ex 3, p. 709		**5**	$x > 13/7$ [See graph above.]	Ex 8, p. 712	
2			Sec 9.7, 1/5	**6**	$\{x \mid x > 4\}$		Sec 9.7, 4/5
3	$\{x \le -10\}$	Ex 4, p. 709		**7**	All numbers less than 11 make this statement true.	Ex 10, p. 712-713	
4	a. $\{x \mid x \ge -5\}$ b. $\{x \mid x \ge -2\}$		Sec 9.7, 2 - 3/5	**8**	All numbers greater than -10 make this statement true.		Sec 9.7, 5/5

☐ **Next, insert your homework.** Make sure you attempt all exercises asked of you and show all work, as in the exercises above. Check your answers if possible. Clearly mark any exercises you were unable to correctly complete so that you may ask questions later. DO NOT ERASE YOUR INCORRECT WORK. THIS IS HOW WE UNDERSTAND AND EXPLAIN TO YOU YOUR ERRORS.

Preparing for the Chapter 9 Test

Start preparing for your Chapter 9 Test as soon as possible. Pay careful attention to any instructor discussion about this test, especially discussion on what sections you will be responsible for, etc.

☐   Work the Chapter 9 Vocabulary Check on page 719.

☐   Read both columns (Definitions and Concepts, and Examples) of the Chapter 9 Highlights starting on page 719.

☐   Read your Class Notes/Examples for each section covered on your Chapter 9 Test. Look for any unresolved questions you may have.

☐   Complete as many of the Chapter 9 Review exercises as possible (starting on page 723). Remember, the odd answers are in the back of your text.

☐   **Most important:** Place yourself in "test" conditions (see below) and work the Chapter 9 Test (page 727 - 728) as a practice test the day before your actual test. To honestly assess how you are doing, try the following:

- Work on a few blank sheets of paper.
- Give yourself the same amount of time you will be given for your actual test.
- Complete this Chapter 9 Practice Test without using your notes or your text.
- If you have any time left after completing this practice test, check your work and try to find any errors on your own.
- Once done, use the back of your book to check ALL answers.
- Try to correct any errors on your own.
- Use the Chapter Test Prep Video (CTPV) to correct any errors you were unable to correct on your own. You can find these videos in the Interactive DVD Lecture Series, in MyMathLab, and on YouTube. Search MartinGayDevMath and click "Channels."

I wish you the best of luck....Elayn Martin-Gay

Section 10.1 The Rectangular Coordinate System

Before Class:

☐   Read the objectives on page 733.

☐   Read the top paragraph and the **Helpful Hint** boxes on page 734.

☐   Complete the exercises:

1. The first number, or *x*-coordinate, of an ordered pair is associated with the *x*-axis. It tells how many units to move _____ or _____ .

2. The second number, or *y*-coordinate, tells how many units to move _____ or _____ .

3. Sketch a rectangular coordinate system. Label the following on your sketch: Quadrants I, II, III, and IV; Origin; *x* – axis; *y* – axis.

During Class:

☐   **Write your class notes.** Neatly write down **all** examples shown as well as key terms or phrases with definitions. If not applicable or if you were absent, watch the Lecture Series (DVD) for this section and do the same (write down the examples shown as well as key terms or phrases). Insert more paper as needed.

Class Notes/Examples	**Your Notes**

Answers: **1)**   left or right   **2)**   up or down   **3)**

Section 10.1 The Rectangular Coordinate System

Class Notes (continued)	**Your Notes**

(Insert additional paper as needed.)

Section 10.1 The Rectangular Coordinate System

Practice:

☐   Complete the Vocabulary and Readiness Check on page 739.

☐   Next, complete any incomplete exercises below. Check and correct your work using the answers and references at the end of this section.

Review this example:

1. Plot each ordered pair. State in which quadrant or on which axis each point lies.

     a. $(5,3)$ b. $(-2,-4)$ c. $(1,-2)$

     d. $(-5,3)$ e. $(0,0)$ f. $(0,2)$

     g. $(-5,0)$ h. $\left(0,-5\frac{1}{2}\right)$ i. $\left(4\frac{2}{3},-3\right)$

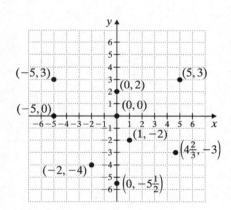

a. Point (5, 3) lies in quadrant I.

b. Point $(-2,-4)$ lies in quadrant III.

c. Point $(1,-2)$ lies in quadrant IV.

d. Point $(-5,3)$ lies in quadrant II.

e. Point (0,0) lies on both the x-axis and y-axis.

f. Point (0, 2) lies on the y-axis.

g. Point $(-5,0)$ lies on the x-axis.

h. Point $\left(0,-5\frac{1}{2}\right)$ lies on the y-axis.

i. Point $\left(4\frac{2}{3},-3\right)$ lies in quadrant IV.

Your turn:

2. Plot each ordered pair. State in which quadrant or on which axis each point lies.

     a. $(1,5)$ b. $(-5,-2)$ c. $(-3,0)$

     d. $(0,-1)$ e. $(2,-4)$ f. $\left(-1,4\frac{1}{2}\right)$

     g. $(3.7,2.2)$ h. $\left(\frac{1}{2},-3\right)$

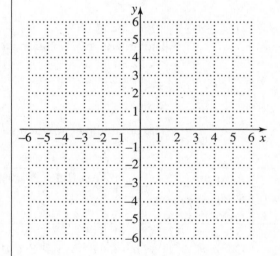

Section 10.1 The Rectangular Coordinate System

Complete this example:

3. Complete the ordered pair so that it is a solution of the equation $3x + y = 12$.

(, 6)

$3x + y = 12$	Original equation
$3x + 6 = 12$	Replace y with 6.
$3x = 6$	Subtract 6 from both sides.
$x = \underline{}$	Divide both sides by 3.

The ordered pair solution is (, 6).

Your turn:

4. Complete the ordered pair so that it is a solution of the equation $x - 4y = 4$.

(, −2), (4,)

Complete this example:

5. Complete the table of ordered pairs for the equation: $y = \dfrac{1}{2}x - 5$

	x	y
a.	−2	
b.	0	

$y = \dfrac{1}{2}x - 5$	Original equation.
$y = \dfrac{1}{2}(-2) - 5$	Let $x = -2$.
$y = \underline{}$	Simplify.
$y = \underline{}$	The ordered pair solution is ◯
$y = \dfrac{1}{2}(0) - 5$	Original equation, with $x = 0$.
$y = \underline{}$	Simplify.
$y = \underline{}$	The ordered pair solution is ◯

Your turn:

6. Complete the table of ordered pairs for the equation: $x = -5y$

x	y
	0
	1
10	

	Answer	Text Ref	Video Ref		Answer	Text Ref	Video Ref
1	See answers on previous page.	Ex 1, p. 734		**4**	(−4,−2), (4, 0)		Sec 10.1, 8/9
2	Q1: a , g Q2: f; Q3: b Q4: e, h x-axis: c y-axis: d		Sec 10.1, 1 - 6/9	**5**	(−2,−6), (0,−5)	Ex 5a–b, p. 737	
3	(2,6)	Ex 3b, p. 736		**6**	(0,0) , (−5,1), (10,−2)		Sec 10.1, 9/9

☐ **Next, insert your homework.** Make sure you attempt all exercises asked of you and show all work, as in the exercises above. Check your answers if possible. Clearly mark any exercises you were unable to correctly complete so that you may ask questions later. DO NOT ERASE YOUR INCORRECT WORK. THIS IS HOW WE UNDERSTAND AND EXPLAIN TO YOU YOUR ERRORS.

Section 10.2 Graphing Linear Equations

Before Class:

☐　Read the objectives on page 745.

☐　Read the **Helpful Hint** boxes on page 746, 748, and 749.

☐　Use the graph below to complete the exercises:

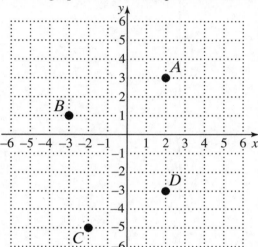

Find the *x*-and *y*-coordinates of each labeled point.

1. Coordinates for point A: _____

2. Coordinates for point B: _____

3. Coordinates for point C: _____

4. Coordinates for point D: _____

During Class:

☐　**Write your class notes.**　Neatly write down **all** examples shown as well as key terms or phrases with definitions. If not applicable or if you were absent, watch the Lecture Series (DVD/CD) for this section and do the same (write down the examples shown as well as key terms or phrases). Insert more paper as needed.

Class Notes/Examples	Your Notes

Answers:　**1)**　(2,3)　　**2)**　(−3,1)　　**3)**　(−2,−5)　　**4)**　(2,−3)

Section 10.2 Graphing Linear Equations

Class Notes (continued)	**Your Notes**

(Insert additional paper as needed.)

Section 10.2 Graphing Linear Equations

Practice:

☐   Next, complete any incomplete exercises below. Check and correct your work using the answers and references at the end of this section.

Review this example:
1. Graph: $-5x + 3y = 15$

Find any three ordered pair solutions. For each solution, choose a value for x or y. Then, place this value in the equation to solve for the other letter.

Let $x = 0$.

$$-5x + 3y = 15$$
$$-5(0) + 3y = 15$$
$$3y = 15$$
$$y = 5$$

Let $x = -2$.

$$-5x + 3y = 15$$
$$-5(-2) + 3y = 15$$
$$10 + 3y = 15$$
$$3y = 5$$
$$y = \frac{5}{3}$$

Let $y = 0$.

$$-5x + 3y = 15$$
$$-5x + 3(0) = 15$$
$$-5x = 15$$
$$x = -3$$

x	y
0	5
-3	0
-2	$\frac{5}{3}$

The graph of $-5x + 3y = 15$ is the line passing through the three points:

$(0,5), (-3,0),$ and $\left(-2, \frac{5}{3}\right)$.

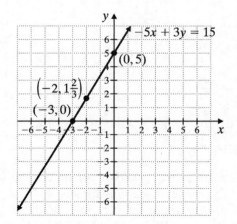

Your turn:
2. Graph: $x - 2y = 6$

x	y

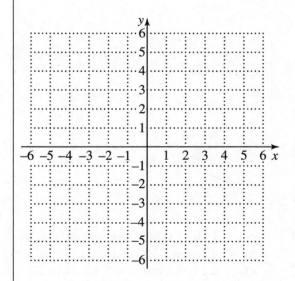

Section 10.2 Graphing Linear Equations

Review this example:	**Your turn:**

3. Graph: $y = -\dfrac{1}{3}x + 2$

Choose three x values and find the corresponding y-values.

If $x = 6$, then $y = -\dfrac{1}{3}(6) + 2 = -2 + 2 = 0$.

If $x = 0$, then $y = -\dfrac{1}{3}(0) + 2 = 0 + 2 = 2$.

If $x = -3$, then $y = -\dfrac{1}{3}(-3) + 2 = 1 + 2 = 3$.

Plot the three ordered pair solutions and graph the line.

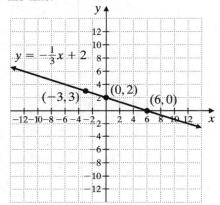

x	y
6	0
0	2
-3	3

4. Graph: $y = \dfrac{1}{2}x + 2$

x	y

	Answer	**Text Ref**	**Video Ref**		**Answer**	**Text Ref**	**Video Ref**
1	See graph on previous page.	Ex 2, p. 747		3	See graph above.	Ex 4, p. 748	
2	Possible points:			4	Possible points:		Sec 10.2, 3/3
			Sec 10.2, 1/3				

☐ **Next, insert your homework.** Make sure you attempt all exercises asked of you and show all work, as in the exercises above. Check your answers if possible. Clearly mark any exercises you were unable to correctly complete so that you may ask questions later. DO NOT ERASE YOUR INCORRECT WORK. THIS IS HOW WE UNDERSTAND AND EXPLAIN TO YOU YOUR ERRORS.

258

Before Class:

☐ Read the objectives on page 756.

☐ Read the **Helpful Hint** boxes on pages 756 and 757. Also read the definition boxes on pages 759 and 760 for vertical and horizontal lines.

☐ Complete the exercises:

1. Graph: $y = 2x - 1$

x	y
2	
0	
−1	

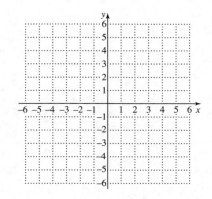

During Class:

☐ **Write your class notes.** Neatly write down **all** examples shown as well as key terms or phrases with definitions. If not applicable or if you were absent, watch the Lecture Series (DVD) for this section and do the same (write down the examples shown as well as key terms or phrases). Insert more paper as needed.

Class Notes/Examples	**Your Notes**

Answers: **1)** $(2,3)$; $(0,-1)$; $(-1,-3)$

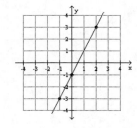

259

Section 10.3 Intercepts

Class Notes (continued)	Your Notes

(Insert additional paper as needed.)

Practice:

☐ Complete the Vocabulary and Readiness Check on page 761.

☐ Next, complete any incomplete exercises below. Check and correct your work using the answers and references at the end of this section.

Review this example:
1. Identify the intercepts.

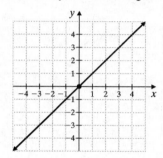

The x- and y-intercept happen to be the same point, (0,0).

Your turn:
2. Identify the intercepts.

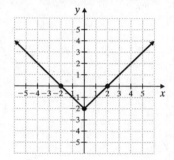

_____

Complete this example:
3. Graph the linear equation by finding and plotting its intercepts: $x - 3y = 6$

Let $y = 0$.   Let $x = 0$.

$x - 3(0) = 6$ $0 - 3y = 6$

$x - 0 = 6$ $-3y = 6$

$x =$ _____ $y =$ _____

The graph of $x - 3y = 6$ is the line through $(6,0)$, $(0,2)$, and $(3,-1)$.

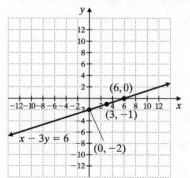

Find a third ordered pair to check your work. Let $x = 3$.

$3 - 3y = 6$

$-3y = 3$

$y = -1$

Your turn:
4. Graph the linear equation by finding and plotting its intercepts:

$$-x + 2y = 6$$

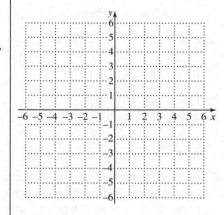

Section 10.3 Intercepts

Complete this example:

5. Graph the linear equation: $x = 2$.

The equation $x = 2$ can be written as $x + 0y = 2$. For any y-value chosen, notice that x is 2.

Graph the ordered pair solutions $(2, 3)$, $(2,0)$, and $(2, -3)$ on the graph below.

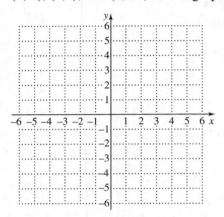

Use a straight-edge to draw the line. The graph is a vertical line.

What is the x-intercept? _____
What is the y-intercept? _____

Your turn:

6. Graph the given linear equations:

   a. $y = 5$ and   b. $x + 3 = 0$

a.

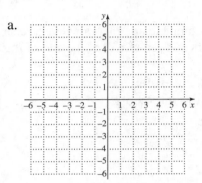

b.

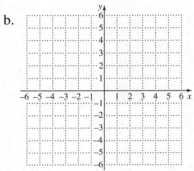

	Answer	Text Ref		Answer	Video Ref
1	$(0,0)$	Ex 3, p. 757	2	$(-2,0)$; $(2,0)$; $(0,-2)$	Sec 10.3, 2/6
3	x–int: $(6,0)$; y–int: $(0,-2)$; see graph on previous page.	Ex 4, p. 757	4		Sec 10.3, 3/6
5	x–int: $(2,0)$; y-int: none	Ex 6, p. 759	6	horizontal line thru $(0,5)$, vertical line thru $(-3,0)$	Sec 10.3, 5 - 6/6

☐   **Next, insert your homework.** Make sure you attempt all exercises asked of you and show all work, as in the exercises above. Check your answers if possible. Clearly mark any exercises you were unable to correctly complete so that you may ask questions later. DO NOT ERASE YOUR INCORRECT WORK. THIS IS HOW WE UNDERSTAND AND EXPLAIN TO YOU YOUR ERRORS.

Section 10.4 Slope and Rate of Change

Before Class:

☐ Read the objectives on page 766.

☐ Read the **Helpful Hint** boxes on pages 766, 767, 768, 770, and 772.

☐ Complete the exercises: Solve each equation for y.

1. $3x + y = 12$ 2. $-3x + y = 12$ 3. $3x + 2y = 12$ 4. $3x - 2y = 12$

During Class:

☐ **Write your class notes.** Neatly write down **all** examples shown as well as key terms or phrases with definitions. If not applicable or if you were absent, watch the Lecture Series (DVD) for this section and do the same (write down the examples shown as well as key terms or phrases). Insert more paper as needed.

Class Notes/Examples	Your Notes

Answers: **1)** $y = -3x + 12$ **2)** $y = 3x + 12$ **3)** $y = -\frac{3}{2}x + 6$ **4)** $y = \frac{3}{2}x - 6$

Section 10.4 Slope and Rate of Change

Class Notes (continued)	**Your Notes**

(Insert additional paper as needed.)

Section 10.4 Slope and Rate of Change

Practice:

☐   Complete the Vocabulary and Readiness Check on page 776.

☐   Next, complete any incomplete exercises below. Check and correct your work using the answers and references at the end of this section.

Complete this example:

1. Find the slope of the line that passes through the points $(-1,5)$ and $(2,-3)$.

Let (x_1, y_1) be $(-1,5)$ and (x_2, y_2) be $(2,-3)$.

$x_1 = $ _____, $y_1 = $ _____
$x_2 = $ _____, $y_2 = $ _____

$m = \dfrac{y_2 - y_1}{x_2 - x_1} = \dfrac{-3-5}{2-(-1)} = \boxed{\dfrac{-8}{3}}$

Your turn:

2. Find the slope of the line that passes through the points $(-1,5)$ and $(6,-2)$.

Complete this example:

3. Use the points shown on the graph to find the slope of the line. The points are $(-1,-2)$ and $(2,4)$.

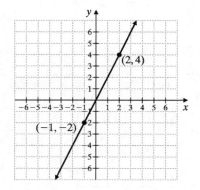

Let (x_1, y_1) be $(2,4)$ and (x_2, y_2) be $(-1,-2)$.

$x_1 = $ _____, $y_1 = $ _____
$x_2 = $ _____, $y_2 = $ _____

$m = \dfrac{y_2 - y_1}{x_2 - x_1} = \dfrac{-2-4}{-1-2} = \dfrac{-6}{-3} = \bigcirc$

Your turn:

4. Use the points shown on the graph to find the slope of the line.

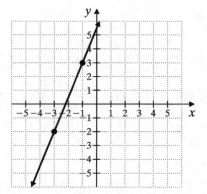

Section 10.4 Slope and Rate of Change

Review this example:	**Your turn:**
5. Find the slope of the line $-2x + 3y = 11$.	**6.** Find the slope of the line:

Review this example:

5. Find the slope of the line $-2x + 3y = 11$.

$-2x + 3y = 11$ Solve the equation for y.

$3y = 2x + 11$

$y = \dfrac{2}{3}x + \dfrac{11}{3}$ The coefficient of x is $\dfrac{2}{3}$.

The slope is $\dfrac{2}{3}$.

Your turn:

6. Find the slope of the line:

$$2x - 3y = 10$$

Review this example:

7. Determine whether the pair of lines is parallel, perpendicular, or neither.

$3x + y = 5$

$2x + 3y = 6$

Solve each equation for y to find each slope.

1st equation: $y = -3x + 5$ Slope is -3.

2nd equation: $2x + 3y = 6$

$\qquad\qquad\qquad 3y = -2x + 6$

$\qquad\qquad\qquad y = -\dfrac{2}{3}x + 2$ Slope is $-\dfrac{2}{3}$.

The slopes are not the same, and their product is not -1.

The lines are neither parallel nor perpendicular.

Your turn:

8. Determine whether the pair of lines is parallel, perpendicular, or neither.

$10 + 3x = 5y$

$5x + 3y = 1$

Hint: Solve each equation for y. Then, compare their slopes.

	Answer	Text Ref	Video Ref		Answer	Text Ref	Video Ref
1	$m = -\dfrac{8}{3}$	Ex 1, p. 767		**5**	$m = \dfrac{2}{3}$	Ex 3, p. 769	
2	$m = -1$		Sec 10.4, 2/11	**6**	$m = \dfrac{2}{3}$		Sec 10.4, 6/11
3	$m = 2$	Ex 2, p. 768		**7**	neither	Ex 7c, p. 772	
4	$m = \dfrac{5}{2}$		Sec 10.4, 1/11	**8**	perpendicular		Sec 10.4, 10/11

☐   **Next, insert your homework.** Make sure you attempt all exercises asked of you and show all work, as in the exercises above. Check your answers if possible. Clearly mark any exercises you were unable to correctly complete so that you may ask questions later. DO NOT ERASE YOUR INCORRECT WORK. THIS IS HOW WE UNDERSTAND AND EXPLAIN TO YOU YOUR ERRORS.

Section 10.5 Equations of Lines

Before Class:

☐   Read the objectives on page 783.

☐   Read the **Helpful Hint** boxes on pages 784 and 786.

☐   Read the Forms of Linear Equations box on page 787.

☐   Complete the exercises:

1. True or False   Two distinct horizontal lines are always parallel.

2. True or False   A horizontal line and a vertical line are always perpendicular.

3. True or False   The slope of a horizontal line is undefined.

4. True or False   The lines $y = \frac{1}{2}x - 2$ and $y = -2x + 3$ are perpendicular.

5. True or False   The lines $y = \frac{1}{2}x - 2$ and $x - 2y = -6$ are parallel.

During Class:

☐   **Write your class notes.** Neatly write down **all** examples shown as well as key terms or phrases with definitions. If not applicable or if you were absent, watch the Lecture Series (DVD) for this section and do the same (write down the examples shown as well as key terms or phrases). Insert more paper as needed.

Class Notes/Examples	**Your Notes**

Answers: **1)** true   **2)** true   **3)** false   **4)** true   **5)** true

Section 10.5 Equations of Lines

Class Notes (continued)	Your Notes

(Insert additional paper as needed.)

Section 10.5 Equations of Lines

Practice:

☐   Complete the Vocabulary and Readiness Check on page 789.

☐   Next, complete any incomplete exercises below. Check and correct your work using the answers and references at the end of this section.

Review this example:	**Your turn:**
1. Write an equation of the line with y-intercept $(0,-3)$ and slope of $\frac{1}{4}$.	**2.** Write an equation of the line with a slope of -4 and y-intercept of $-\frac{1}{6}$.

We are given the slope and the y-intercept.
We let $m=\frac{1}{4}$ and $b=-3$. We will write the equation in slope-intercept form.

$y = mx + b$

$y = \frac{1}{4}x + (-3)$ Let $m=\frac{1}{4}$ and $b=-3$.

$\boxed{y = \frac{1}{4}x - 3}$ Simplify.

Review this example:	**Your turn:**
3. Use the slope-intercept form to graph the equation $y = \frac{3}{5}x - 2$.	**4.** Use the slope-intercept form to graph the equation $y = -5x$.

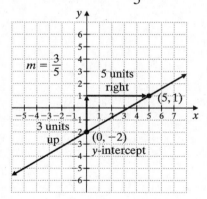

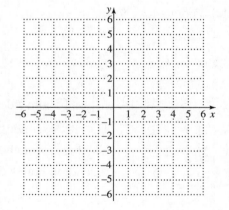

$m = \frac{3}{5} \; \frac{rise}{run}$ and $b = -2$.

First, plot $(0,-2)$.

From there move up 3 and right 5.

269

Section 10.5 Equations of Lines

Review this example:

5. Find an equation of the line with slope -2 that passes through $(-1,5)$. Write the equation in standard form, $Ax+By=C$.

We are given the slope and a point on the line.
Use the point-slope form: $y-y_1=m(x-x_1)$

$y-5=-2[x-(-1)]$ $m=-2$, $(x_1,y_1)=(-1,5)$

$y-5=-2(x+1)$ Simplify.

$y-5=-2x-2$ Distribute.

$y=-2x+3$ Add 5 to both sides.

$\boxed{2x+y=3}$ Add $2x$ to both sides.

Your turn:

6. Find an equation of the line with slope $m=-8$ that passes through the point $(-1,-5)$. Write the equation in the form $Ax+By=C$.

Review this example:

7. Find an equation of the line through $(2,5)$ and $(-3,4)$. Write the equation in the form $Ax+By=C$.

First, use the two given points to find the slope of

the line. $m=\dfrac{4-5}{-3-2}=\dfrac{-1}{-5}=\dfrac{1}{5}$

We will use point-slope form, $y-y_1=m(x-x_1)$.

$y-5=\dfrac{1}{5}(x-2)$ Let $m=\dfrac{1}{5}$. Use $(2,5)$.

$5(y-5)=5\cdot\dfrac{1}{5}(x-2)$ Multiply both sides by 5.

$5y-25=x-2$ Distribute, simplify.

$-x+5y-25=-2$ Subtract x from both sides.

$\boxed{-x+5y=23}$ Add 25 to both sides.

Your turn:

8. Find an equation of the line through $(2,3)$ and $(-1,-1)$. Write the equation in the form $Ax+By=C$.

Section 10.5 Equations of Lines

	Answer	Text Ref	Video Ref		Answer	Text Ref	Video Ref
1	$y = \dfrac{1}{4}x - 3$	Ex 3 p. 784		5	$2x + y = 3$	Ex 4 p. 785	
2	$y = -4x - \dfrac{1}{6}$		Sec 10.5, 1/8	6	$8x + y = -13$		Sec 10.5, 4/8
3	See graph on previous page.	Ex 1 p. 783		7	$-x + 5y = 23$	Ex 5 p. 785	
4			Sec 10.5, 2/8	8	$4x - 3y = -1$		Sec 10.5, 5/8

☐ **Next, insert your homework.** Make sure you attempt all exercises asked of you and show all work, as in the exercises above. Check your answers if possible. Clearly mark any exercises you were unable to correctly complete so that you may ask questions later. DO NOT ERASE YOUR INCORRECT WORK. THIS IS HOW WE UNDERSTAND AND EXPLAIN TO YOU YOUR ERRORS.

Section 10.6 Introduction to Functions

Before Class:

☐   Read the objectives on page 796.

☐   Read the **Helpful Hint** boxes on page 801.

☐   Complete the exercises:

Use $m = \dfrac{y_2 - y_1}{x_2 - x_1}$ to find the slope of the line that passes through the given points.

1. $(3, -1)$ and $(-2, -5)$ 2. $(-2, -1)$ and $(-2, -5)$

During Class:

☐   **Write your class notes.** Neatly write down **all** examples shown as well as key terms or phrases with definitions. If not applicable or if you were absent, watch the Lecture Series (DVD) for this section and do the same (write down the examples shown as well as key terms or phrases). Insert more paper as needed.

Class Notes/Examples	**Your Notes**

Answers: **1)** $m = \dfrac{4}{5}$ **2)** undefined slope

Section 10.6 Introduction to Functions

Class Notes (continued)	Your Notes

(Insert additional paper as needed.)

Practice:

☐   Complete the Vocabulary and Readiness Check on page 803.

☐   Next, complete any incomplete exercises below. Check and correct your work using the answers and references at the end of this section.

Review this example:

1. Find the domain and range of the relation.

$$\{(0,2),(3,3),(-1,0),(3,-2)\}$$

The domain is the set of all x-coordinates and the range is the set of all y-coordinates.

Domain: $\{-1,0,3\}$ Range: $\{-2,0,2,3\}$

Your turn:

2. Find the domain and range of the relation.

$$\{(0,-2),(1,-2),(5,-2)\}$$

Review this example:

3. Which of the following relations are also functions?

    a. $\{(-1,1),(2,3),(7,3),(8,6)\}$

    b. $\{(0,-2),(1,5),(0,3),(7,7)\}$

A function cannot have two ordered pairs with the same x-coordinate but different y-coordinate.

a. The x-coordinate -1 is paired with 1.
    2 is paired with 3.
    7 is paired with 3.
    8 is paired with 6.

Since each x-coordinate is paired with only one y-value, the relation is a function.

b. The x-coordinate 0 is paired with -2.
    1 is paired with 5.
    0 is paired with 3.
    7 is paired with 7.

Since the x-coordinate 0 is paired with **two y-values**, -2 and 3, the relation is not a function.

Your turn:

4. Determine whether the relation $\{(-1,0),(-1,6),(-1,8)\}$ is also a function.

275

Section 10.6 Introduction to Functions

Complete this example:

5. Is the given graph the graph of a function?

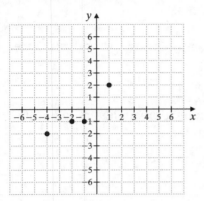

The relation consists of the points:

$$\{(-4,-2),(-2,-1),(-1,-1),(1,2)\}$$

Each x-coordinate is paired with exactly ⬭ y-coordinate.

Is the graph a function?   Yes   or   No

Your turn:

6. Is the given graph the graph of a function?

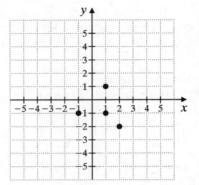

Review this example:

7. For $g(x) = x^2 - 3$ find $g(2), g(-2),$ and $g(0)$.

$$g(2) = (2)^2 - 3 = 4 - 3 = \boxed{1}$$

$$g(-2) = (-2)^2 - 3 = 4 - 3 = \boxed{1}$$

$$g(0) = (0)^2 - 3 = 0 - 3 = \boxed{-3}$$

Your turn:

8. For $f(x) = x^2 + 2$ find $f(-2),$ $f(0),$ and $f(3)$.

	Answer	Text Ref	Video Ref		Answer	Video Ref	Video Ref
1	D: $\{-1,0,3\}$ R: $\{-2,0,2,3\}$	Ex 1, p. 796		5	one; yes a function	Ex 3a, p. 797	
2	D: $\{0,1,5\}$ R: $\{-2\}$		Sec 10.6, 1/13	6	not a function		Sec 10.6, 4/13
3	a. yes a function b. not a function	Ex 2a, b, p. 797		7	1, 1, −3	Ex 7, p. 801	
4	not a function		Sec 10.6, 3/13	8	6, 2, 11		Sec 10.6, 10/13

☐   **Next, insert your homework.** Make sure you attempt all exercises asked of you and show all work, as in the exercises above. Check your answers if possible. Clearly mark any exercises you were unable to correctly complete so that you may ask questions later. DO NOT ERASE YOUR INCORRECT WORK. THIS IS HOW WE UNDERSTAND AND EXPLAIN TO YOU YOUR ERRORS.

Section 10.7 Graphing Linear Inequalities in Two Variables

Before Class:

☐ Read the objectives on page 808.

☐ Read the **Helpful Hint** box on page 811.

☐ Complete the exercises:

Find the domain and the range of each function graphed.

1.

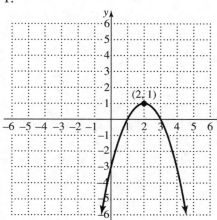

2.

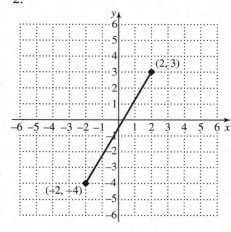

Domain: _____
Range: _____

Domain: _____
Range: _____

During Class:

☐ **Write your class notes.** Neatly write down **all** examples shown as well as key terms or phrases with definitions. If not applicable or if you were absent, watch the Lecture Series (DVD) for this section and do the same (write down the examples shown as well as key terms or phrases). Insert more paper as needed.

Class Notes/Examples   **Your Notes**

Answers: **1)** Domain: $-2 \leq x \leq 2$, Range: $-4 \leq y \leq 3$ **2)** Domain: All real numbers, Range: $y \leq 1$

277

Section 10.7 Graphing Linear Inequalities in Two Variables

Class Notes (continued)	Your Notes

(Insert additional paper as needed.)

Section 10.7 Graphing Linear Inequalities in Two Variables

Practice:

☐   Complete the Vocabulary and Readiness Check on page 814.

☐   Next, complete any incomplete exercises below. Check and correct your work using the answers and references at the end of this section.

Review this example:	**Your turn:**
1. Determine whether the ordered pair is a solution of the inequality $2x - y < 6$.	**2.** Determine whether each ordered pair is a solution of the inequality $x < -y$.

Review this example:

1. Determine whether the ordered pair is a solution of the inequality $2x - y < 6$.

$(5, -1)$

We replace x with 5 and y with -1 and see if a true statement results.

$$2x - y < 6$$
$$2(5) - (-1) < 6 \quad \text{Replace } x \text{ with } 5; \ y \text{ with } -1.$$
$$10 + 1 < 6 \quad \text{Simplify.}$$
$$11 < 6 \quad \text{False}$$

The ordered pair is not a solution.   ⟨No⟩

Your turn:

2. Determine whether each ordered pair is a solution of the inequality $x < -y$.

a. $(0, 2)$ b. $(-5, 1)$

Complete this example:

3. Graph the inequality: $x + y < 7$

Step 1. Find 2 or 3 points on the line.

x	y
0	
2	
7	

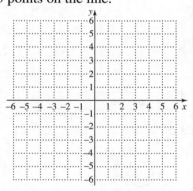

Plot these points on the above graph.
Connect this boundary line with a dashed line because the inequality sign is $<$.

Step 2. Choose a test point; be careful *not* to choose a point on the boundary line.

(*continued on next page*)

Your turn:

4. Graph the inequality: $y \geq 2x$

Step 1. Find 2 or 3 points on the line.

x	y
0	
1	
2	

Step 2. Choose a test point; be careful *not* to choose a point on the boundary line.

Step 3. Graph – see next page.

Section 10.7 Graphing Linear Inequalities in Two Variables

Choose $(0,0)$. Substitute into $x+y<7$.

$0+0<7$ Replace x with 0 and y with 0.

    $0<7$ True

Step 3. Since the result is a true statement, $(0,0)$ is a solution of $x+y<7$.

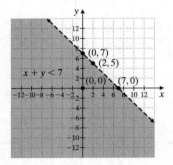

Every point in the same half-plane as $(0,0)$ is also a solution.

Shade the half-plane containing $(0, 0)$, as shown.

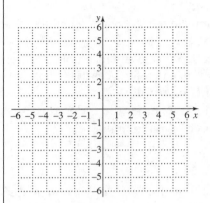

Review this example:

5. Graph the inequality: $5x+4y\le 20$

Graph the solid boundary line, $5x+4y=20$.

Test $(0,0)$.

    $5x+4y\le 20$

    $5(0)+4(0)\le 20$

        $0\le 20$ True

Shade the half-plane that contains $(0,0)$.

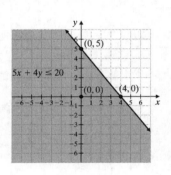

Your turn:

6. Graph the inequality: $2x+7y>5$

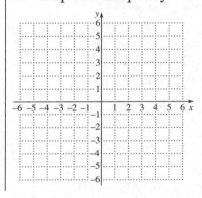

	Answer	Text Ref		Answer	Video Ref
1	no	Ex 1a, p. 808	2	a. no b. yes	Sec 10.7, 1/3
3	See graph above.	Ex 2, p. 810	4		Sec 10.7, 2/3
5	See graph above.	Ex 5, p. 812	6		Sec 10.7, 3/3

☐   **Next, insert your homework.** Make sure you attempt all exercises asked of you and show all work, as in the exercises above. Check your answers if possible. Clearly mark any exercises you were unable to correctly complete so that you may ask questions later. DO NOT ERASE YOUR INCORRECT WORK. THIS IS HOW WE UNDERSTAND AND EXPLAIN TO YOU YOUR ERRORS.

Section 10.8 Direct and Inverse Variation

Before Class:

☐   Read the objectives on page 818.

☐   Read the **Helpful Hint** box on page 822.

☐   Complete the exercises:

    1.   In the equation $y = kx$, let $y = 24$ and $x = 12$. Solve for k.

    2.   In the equation $y = \dfrac{k}{x}$, let $y = 24$ and $x = 12$. Solve for k.

During Class:

☐   **Write your class notes.** Neatly write down **all** examples shown as well as key terms or phrases with definitions. If not applicable or if you were absent, watch the Lecture Series (DVD) for this section and do the same (write down the examples shown as well as key terms or phrases). Insert more paper as needed.

<table>
<tr><th>Class Notes/Examples</th><th>Your Notes</th></tr>
</table>

Answers: **1)** $k = 2$ **2)** $k = 288$

Section 10.8 Direct and Inverse Variation

Class Notes (continued)	Your Notes

(Insert additional paper as needed.)

Practice:

☐   Complete the Vocabulary and Readiness Check on page 825.

☐   Next, complete any incomplete exercises below. Check and correct your work using the answers and references at the end of this section.

Review this example:

1. Write a direct variation equation, $y = kx$, that satisfies the ordered pairs in the table.

x	2	9	1.5	−1
y	6	27	4.5	−3

To find k choose one ordered pair in the table and substitute the values into the equation. If we choose $(2,6)$, let $x = 2$ and $y = 6$. Solve for **k**.

$$6 = k \cdot 2$$
$$\frac{6}{2} = \frac{k \cdot 2}{2}$$
$$3 = k$$

The direct variation equation is $y = 3x$.

Your turn:

2. Write a direct variation equation, $y = kx$, that satisfies the ordered pairs in the table.

x	−2	2	4	5
y	−12	12	24	30

Review this example:

3. Write an inverse variation equation, $y = \dfrac{k}{x}$, that satisfies the ordered pairs in the table.

x	2	4	$\dfrac{1}{2}$
y	6	3	24

Choose $(2,6)$. Then, let $x = 2$ and $y = 6$.

$$6 = \frac{k}{2}$$
$$2(6) = 2 \cdot \frac{k}{2} \qquad \text{Multiply both sides by 2.}$$
$$12 = k \qquad \text{Solve for } k.$$

The inverse variation equation is $y = \dfrac{12}{x}$.

Your turn:

4. Write an inverse variation equation, $y = \dfrac{k}{x}$, that satisfies the ordered pairs in the table.

x	1	−7	3.5	−2
y	7	−1	2	−3.5

283

Section 10.8 Direct and Inverse Variation

Complete this example:

5. Suppose that y varies inversely as x. If $y = 0.02$ when $x = 75$, find the constant of variation and the inverse variation equation. Then find y when x is 30.

Find k, the constant of variation:

Use the inverse variation equation: $y = \dfrac{k}{x}$

Let $y = 0.02$ and $x = 75$.

$$0.02 = \frac{k}{75}$$

$$75(0.02) = 75 \cdot \frac{k}{75} \qquad \text{Multiply both sides by 75.}$$

$$\boxed{1.5 = k} \qquad \text{Solve for } k.$$

Find the inverse variation equation:

The inverse variation equation is ⬯

Find y when x = 30:

$$y = \frac{1.5}{30} = \bigcirc \qquad \text{Replace } x \text{ with 30.}$$

Thus, when x is 30, y is ⬯.

Your turn:

6. y varies inversely as x. If $y = 5$ when $x = 60$, find y when x is 100.

	Answer	Text Ref	Text Ref	Answer	Text Ref	Video Ref
1	$y = 3x$	Ex 1, p. 819		4 $\quad y = \dfrac{7}{x}$		Sec 10.8, 3/8
2	$y = 6x$		Sec 10.8, 1/8	5 $\quad y = \dfrac{1.5}{x}; 0.05$ When x is 30, y is 0.05.	Ex 5, p. 822	
3	$y = \dfrac{12}{x}$	Ex 4, p. 821		6 $\quad y = 3$		Sec 10.8, 4/8

☐ **Next, insert your homework.** Make sure you attempt all exercises asked of you and show all work, as in the exercises above. Check your answers if possible. Clearly mark any exercises you were unable to correctly complete so that you may ask questions later. DO NOT ERASE YOUR INCORRECT WORK. THIS IS HOW WE UNDERSTAND AND EXPLAIN TO YOU YOUR ERRORS.

Preparing for the Chapter 10 Test

Start preparing for your Chapter 10 Test as soon as possible. Pay careful attention to any instructor discussion about this test, especially discussion on what sections you will be responsible for, etc.

☐   Work the Chapter 10 Vocabulary Check on page 829.

☐   Read both columns (Definitions and Concepts, and Examples) of the Chapter 10 Highlights starting on page 829.

☐   Read your Class Notes/Examples for each section covered on your Chapter 10 Test. Look for any unresolved questions you may have.

☐   Complete as many of the Chapter 10 Review exercises as possible (page 834). Remember, the odd answers are in the back of your text.

☐   **Most important:** Place yourself in "test" conditions (see below) and work the Chapter 10 Test (pages 840 - 842) as a practice test the day before your actual test. To honestly assess how you are doing, try the following:

- Work on a few blank sheets of paper.
- Give yourself the same amount of time you will be given for your actual test.
- Complete this Chapter 10 Practice Test without using your notes or your text.
- If you have any time left after completing this practice test, check your work and try to find any errors on your own.
- Once done, use the back of your book to check ALL answers.
- Try to correct any errors on your own.
- Use the Chapter Test Prep Video (CTPV) to correct any errors you were unable to correct on your own. You can find these videos in the Interactive DVD Lecture Series, in MyMathLab, and on YouTube. Search Martin-Gay Prealgebra & Introductory Algebra and click "Channels."

I wish you the best of luck….Elayn Martin-Gay

Section 11.1 Solving Systems of Linear Equations by Graphing

Before Class:

☐ Read the objectives on page 847.

☐ Read the **Helpful Hint** boxes on page 848.

☐ Complete the exercises:

1. Determine whether the following lines are parallel, perpendicular or neither.

a. $\begin{cases} y = 2x + 5 \\ y = -2x + 3 \end{cases}$

b. $\begin{cases} y = -2x - 9 \\ 2x + y = 11 \end{cases}$

c. $\begin{cases} x = 3 \\ y = 4 \end{cases}$

d. $\begin{cases} -2x + 3y = 1 \\ 3x + 2y = 12 \end{cases}$

During Class:

☐ **Write your class notes.** Neatly write down **all** examples shown as well as key terms or phrases with definitions. If not applicable or if you were absent, watch the Lecture Series (DVD) for this section and do the same (write down the examples shown as well as key terms or phrases). Insert more paper as needed.

Class Notes/Examples	**Your Notes**

Answers: **1)** a. neither   b. parallel   c. perpendicular   d. perpendicular

Section 11.1 Solving Systems of Linear Equations by Graphing

Class Notes (continued)	**Your Notes**

(Insert additional paper as needed.)

Section 11.1 Solving Systems of Linear Equations by Graphing

Practice:

☐   Complete the Vocabulary and Readiness Check on page 852.

☐   Next, complete any incomplete exercises below. Check and correct your work using the answers and references at the end of this section.

Review this example:

1. Determine whether $(-1, 2)$ is a solution of

the system: $\begin{cases} x + 2y = 3 \\ 4x - y = 6 \end{cases}$

Replace x with -1 and y with 2.

$x + 2y = 3$ $\qquad\qquad$ $4x - y = 6$

$-1 + 2(2) \overset{?}{=} 3$ $\qquad$ $4(-1) - 2 \overset{?}{=} 6$

$-1 + 4 \overset{?}{=} 3$ $\qquad\qquad$ $-4 - 2 \overset{?}{=} 6$

$\qquad 3 \overset{?}{=} 3$ **True** $\qquad -6 \overset{?}{=} 6$ **False**

$(-1, 2)$ is not a solution of the system.

Your turn:

2. Determine whether each ordered pair is a solution of the system of linear equations.

$\begin{cases} 3x - y = 5 \\ x + 2y = 11 \end{cases}$

a. $(3, 4)$

b. $(0, -5)$

Complete this example:

3. Solve this system of linear equations by

graphing: $\begin{cases} 2x + 3y = -2 \\ x = 2 \end{cases}$

$2x + 3y = -2$	
x	y
0	
2	

$x = 2$	
x	y
	0
	2

Find two ordered pair solutions.

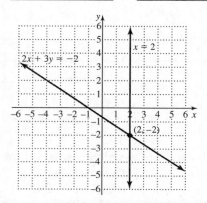

The two lines appear to intersect at $(2, -2)$.

Your turn:

4. Solve the system of linear equations by

graphing: $\begin{cases} 2x + y = 0 \\ 3x + y = 1 \end{cases}$

Find two ordered pair solutions for $2x + y = 0$ and $3x + y = 1$.

x	y
0	
2	

x	y
0	
2	

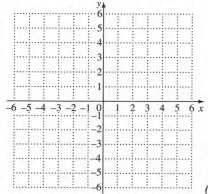

(continued)

289

Section 11.1 Solving Systems of Linear Equations by Graphing

(continued)

To check, replace x with 2 and y with -2:

$2x + 3y = -2$ 1st equation | $x = 2$ 2nd equation

$2(2) + 3(-2) \overset{?}{=} -2$ | $2 \overset{?}{=} 2$

$4 + (-6) \overset{?}{=} -2$ | $2 = 2$ **True**

$-2 = -2$ **True**

$(2, -2)$ is the solution of the system.

The two lines appear to intersect at _____.

Check:

$2 \overset{?}{=} 2$

$2 = 2$ **True**

Review this example:

5. Without graphing, determine the number of solutions for:
$$\begin{cases} \dfrac{1}{2}x - y = 2 \\ x = 2y + 5 \end{cases}$$

$\dfrac{1}{2}x - y = 2$ **First equation**

$\dfrac{1}{2}x = y + 2$ Add y to both sides.

$\dfrac{1}{2}x - 2 = y$ Subtract 2 from both sides.

Slope is $\dfrac{1}{2}$; y-intercept is -2.

$x = 2y + 5$ **Second equation**

$x - 5 = 2y$ Subtract 5 from both sides.

$\dfrac{x}{2} - \dfrac{5}{2} = \dfrac{2y}{2}$ Divide both sides by 2.

$\dfrac{1}{2}x - \dfrac{5}{2} = y$ Slope is $\dfrac{1}{2}$; y-intercept is $-\dfrac{5}{2}$.

Lines are parallel, no solution.

Your turn:

6. Without graphing, determine the number of solutions of the system.
$$\begin{cases} 4x + y = 24 \\ x + 2y = 2 \end{cases}$$

	Answer	Text Ref	Video Ref		Answer	Text Ref	Video Ref
1	no	Ex 2, p. 847		4			Sec 11.1, 2/7
2	a. yes b. no		Sec 11.1, 1/7	5	parallel; no solution	Ex 7, p. 850	
3	$(2,-2)$; see graph above	Ex 4, p. 848-849		6	intersecting; one solution		Sec 11.1, 5/7

☐ **Next, insert your homework.** Make sure you attempt all exercises asked of you and show all work, as in the exercises above. Check your answers if possible. Clearly mark any exercises you were unable to correctly complete so that you may ask questions later. DO NOT ERASE YOUR INCORRECT WORK. THIS IS HOW WE UNDERSTAND AND EXPLAIN TO YOU YOUR ERRORS.

Section 11.2 Solving Systems of Linear Equations By Substitution

Before Class:

☐   Read the objective on page 858.

☐   Read the **Helpful Hint** boxes on pages 858, 860, and 861.

☐   Complete the exercise:

1.   Determine whether the ordered pair $(4, 2)$ is a solution of the system of linear equations.

$$\begin{cases} 2x + y = 10 \\ x = y + 2 \end{cases}$$

During Class:

☐   **Write your class notes.** Neatly write down **all** examples shown as well as key terms or phrases with definitions. If not applicable or if you were absent, watch the Lecture Series (DVD) for this section and do the same (write down the examples shown as well as key terms or phrases.) Insert more paper as needed.

Class Notes/Examples	Your Notes

Answers: **1)**   1st equation: 10 = 10, True; 2nd equation: 4 = 4, True. Yes, the solution is (4, 2).

Section 11.2 Solving Systems of Linear Equations By Substitution

Class Notes (continued)	**Your Notes**

(Insert additional paper as needed.)

Section 11.2 Solving Systems of Linear Equations By Substitution

Practice:

☐   Complete the Vocabulary and Readiness Check on page 863.

☐   Next, complete any incomplete exercises below. Check and correct your work using the answers and references at the end of this section.

Complete this example:

1. Solve the system: $\begin{cases} 5x - y = -2 \\ y = 3x \end{cases}$

The second equation is solved for y in terms of x. Substitute $3x$ for y in the first equation.

$$5x - y = -2 \qquad \text{First equation}$$
$$5x - (3x) = -2 \qquad \text{Substitute } 3x \text{ for } y.$$
$$5x - 3x = -2$$
$$2x = -2 \qquad \text{Combine like terms.}$$
$$x = -1 \qquad \text{Divide both sides by 2.}$$

The x-value of the ordered pair solution is -1.

In the second equation, replace x with -1 to find the corresponding y-value.

$$y = 3x \qquad \text{Second equation}$$
$$y = 3(-1) \qquad \text{Replace } x \text{ with } -1.$$
$$y = -3$$

The y-value of the ordered pair solution is -3.

The solution to the system is $(-1, -3)$.

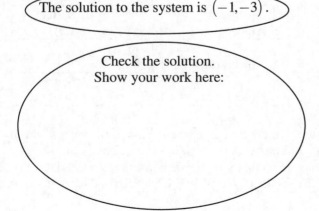

Check the solution.
Show your work here:

Your turn:

2. Solve the system: $\begin{cases} x + y = 6 \\ y = -3x \end{cases}$

Check the solution.
Show your work here:

Section 11.2 Solving Systems of Linear Equations By Substitution

Review this example:

3. Solve the system: $\begin{cases} 6x+12y=5 \\ -4x-8y=0 \end{cases}$

Your turn:

4. Solve the system: $\begin{cases} 3x+6y=9 \\ 4x+8y=16 \end{cases}$

Select one of the equations to solve for either x or y.

$-4x-8y=0$	Second equation
$-4x=8y$	Solve for x.
$\dfrac{-4x}{-4}=\dfrac{8y}{-4}$	Divide both sides by -4.
$x=-2y$	Simplify.

$6x+12y=5$	First equation
$6(-2y)+12y=5$	Replace x with $-2y$
$-12y+12y=5$	Multiply.
$0=5$	Combine like terms.

The statement $0=5$ is false. This indicates the system has **no solution**. Graphically, this means the lines do not intersect and are parallel.

	Answer	Text Ref	Video Ref		Answer	Text Ref	Video Ref
1	Solution: $(-1,-3)$	Ex 2, p. 859		**3**	No solution	Ex 6, p. 862	
2	Solution: $(-3,9)$		Sec 11.2, 1/4	**4**	No solution		Sec 11.2, 3/4

☐   **Next, insert your homework.** Make sure you attempt all exercises asked of you and show all work, as in the exercises above. Check your answers if possible. Clearly mark any exercises you were unable to correctly complete so that you may ask questions later. DO NOT ERASE YOUR INCORRECT WORK. THIS IS HOW WE UNDERSTAND AND EXPLAIN TO YOU YOUR ERRORS.

Section 11.3 Solving Systems of Linear Equations by Addition

Before Class:

☐   Read the objective on page 866.

☐   Read the **Helpful Hint** boxes on pages 866 and 867.

☐   Complete the exercises:

1. Give the solution of each system. If the system has no solution or an infinite number of solutions, say so. If the system has one solution, find it.

a. $\begin{cases} x = 2y \\ x + y = 3 \end{cases}$
          b. $\begin{cases} 2y = x + 2 \\ 6x - 12y = 0 \end{cases}$

  When solving you obtain $x = 2$.
       When solving you obtain $0 = 12$.

c. $\begin{cases} \dfrac{1}{4}x - 2y = 1 \\ x - 8y = 4 \end{cases}$
      d. $\begin{cases} 3y - x = 6 \\ 4x + 12y = 0 \end{cases}$

  When solving you obtain $0 = 0$.
        When solving you obtain $y = 1$.

During Class:

☐   **Write your class notes.** Neatly write down **all** examples shown as well as key terms or phrases with definitions. If not applicable or if you were absent, watch the Lecture Series (DVD) for this section and do the same (write down the examples shown as well as key terms or phrases.) Insert more paper as needed.

Class Notes/Example	Your Notes

Answers: **1)** a. $(2,1)$ b. no solution   c. infinite number of solutions   d. $(-3,1)$

Section 11.3 Solving Systems of Linear Equations by Addition

Class Notes (continued)	Your Notes

(Insert additional paper as needed.)

Section 11.3 Solving Systems of Linear Equations by Addition

Practice:

☐   Complete the Vocabulary and Readiness Check on page 870.

☐   Next, complete any incomplete exercises below. Check and correct your work using the answers and references at the end of this section.

Review this example:

1. Solve the system of equations using the addition method.

$$\begin{cases} x + y = 7 \\ x - y = 5 \end{cases} \quad \text{Add the equations.}$$

$\quad\quad 2x = 12$ Eliminate y.
$\quad\quad\quad x = 6$ Divide both sides by 2.

To find the corresponding y-value, let $x = 6$ in either equation.

$x + y = 7$ Use the first equation.
$6 + y = 7$ Let $x = 6$.
$\quad\quad y = 1$ Solve for y.
The solution is $(6, 1)$.

Check: First equation   Second equation
$\quad\quad\quad x + y = 7 \quad\quad\quad\quad x - y = 5$

$\quad\quad\quad 6 + 1 \overset{?}{=} 7 \quad\quad\quad\quad 6 - 1 \overset{?}{=} 5$
$\quad\quad\quad\quad 7 = 7 \text{ True} \quad\quad\quad 5 = 5 \text{ True}$

Your turn:

2. Solve the system of equations using the addition method.

$$\begin{cases} x - 2y = 8 \\ -x + 5y = -17 \end{cases}$$

Complete this example:

3. Solve the system of equations using the addition method.

$$\begin{cases} -2x + y = 2 \\ -x + 3y = -4 \end{cases}$$

First, make the coefficients of one of the variables opposites, so the sum of the terms is 0.

Use the multiplication property of equality.
$-2x + y = 2$ First equation

$-3(-2x + y) = -3(2)$ Multiply both sides by –3.

$\quad 6x - 3y = -6$ Resulting equation 1

(continued on next page)

Your turn:

4. Solve the system of equations using the addition method.

$$\begin{cases} 3x - 2y = 7 \\ 5x + 4y = 8 \end{cases}$$

(continued on next page)

297

Section 11.3 Solving Systems of Linear Equations by Addition

$$\begin{cases} 6x - 3y = -6 \\ -x + 3y = -4 \end{cases}$$ Add the equations.

$\underline{\hspace{6em}}$

$5x \quad = -10$ Eliminate y.

$x = -2$ Divide both sides by 5.

To find the corresponding y-value, let $x = -2$
in either equation.

$-2x + y = 2$ Use the original first equation.

$-2(\underline{\hspace{1.5em}}) + y = 2$ Let $x = -2$.

$\underline{\hspace{6em}}$ Simplify.

$\underline{\hspace{6em}}$ Solve for y.

The solution is $(\quad , \quad)$.

Check: First equation   Second equation
     $-2x + y = 2$ $-x + 3y = -4$

Check:

Review this example:	**Your turn:**

6. Solve this system of equations using
   the addition method: $\begin{cases} 3x - 2y = 2 \\ -9x + 6y = -6 \end{cases}$

Multiply both sides of the first equation by 3.

$3(3x - 2y) = 3(2)$ The result is $9x - 6y = 6$.

$\begin{cases} 9x - 6y = 6 \\ -9x + 6y = -6 \end{cases}$ Add the equations.

$\underline{\hspace{6em}}$

$0 = 0$ Both variables are eliminated.

These are the same equations and have

an (infinite number of solutions)

7. Solve the system of equations using
   the addition method.

$$\begin{cases} \dfrac{x}{3} - y = 2 \\ -\dfrac{x}{2} + \dfrac{3y}{2} = -3 \end{cases}$$ Hint: Clear fractions!

	Answer	Text Ref	Video Ref		Answer	Text Ref	Video Ref
1	$(6,1)$	Ex 1, p. 866		**4**	$(2, -\frac{1}{2})$		Sec 11.3, 2/4
2	$(2,-3)$		Sec 11.3, 1/4	**5**	Infinite number of solutions	Ex 4, p. 868	
3	$(-2,-2)$	Ex 2, p. 867		**6**	Infinite number of solutions		Sec 11.3, 3/4

☐   **Next, insert your homework.** Make sure you attempt all exercises asked of you and show
all work, as in the exercises above. Check your answers if possible. Clearly mark any
exercises you were unable to correctly complete so that you may ask questions later. DO
NOT ERASE YOUR INCORRECT WORK. THIS IS HOW WE UNDERSTAND AND
EXPLAIN TO YOU YOUR ERRORS.

Section 11.4 Systems of Linear Equations and Problem Solving

Before Class:

☐   Read the objective on page 875.

☐   Read the Problem-Solving Steps on page 875.

☐   Complete the exercise:

1.   Solve the system using the addition method: $\begin{cases} x + y = 3 \\ x - y = 7 \end{cases}$

During Class:

☐   **Write your class notes.** Neatly write down **all** examples shown as well as key terms or phrases with definitions. If not applicable or if you were absent, watch the Lecture Series (DVD) for this section and do the same (write down the examples shown as well as key terms or phrases). Insert more paper as needed.

Class Notes/Examples	Your Notes

Answers: **1)** $(5, -2)$

Section 11.4 Solving Systems of Linear Equations and Problem Solving

Class Notes (continued)	**Your Notes**

(Insert additional paper as needed.)

Section 11.4 Systems of Linear Equations and Problem Solving

Practice:

☐   Next, complete any incomplete exercises below. Check and correct your work using the answers and references at the end of this section.

Review this example:
1. Find two numbers whose sum is 37 and whose difference is 21.

UNDERSTAND.

Think of two numbers that add up to 37. Suppose we have $30 + 7 = 37$. What is the difference of the two numbers? $30 - 7 = 23$. The difference is not 21, but we can let $x =$ the first number and $y =$ the second number to determine the actual numbers.

TRANSLATE.

Two numbers whose sum is 37: $x + y = 37$
Two numbers whose difference is 21: $x - y = 21$

SOLVE: $\begin{cases} x + y = 37 \\ x - y = 21 \end{cases}$

The coefficients of y are opposites, so we can use the addition method and add the equations by columns.

$x + y = 37$
$\underline{x - y = 21}$
$\quad 2x = 58$ Add the equations.
$\quad\;\; x = 29$ Divide both sides by 2.

$\;\; x + y = 37$ First equation
$29 + y = 37$ Replace x with 29.
$\qquad y = 8$ Subtract 29 from both sides.

INTERPRET. The solution of the system is $(29, 8)$.

$29 + 8 = 37$
$29 - 8 = 21$ Both equations are true.

The numbers are 29 and 8.

Your turn:
2. Two numbers total 83 and have a difference of 17. Find the two numbers.

301

Section 11.4 Solving Systems of Linear Equations and Problem Solving

Complete this example:

3. The Cirque du Soleil show Varekai is performing locally. Matinee admission for 4 adults and 2 children is $374, while admission for 2 adults and 3 children is $285. What is the price of an adult's ticket and what is the price of a child's ticket?

Let A = the price of an adult's ticket and
C = the price of a child's ticket.

TRANSLATE.

admission for 4 adults	and	admission for 2 children	is	$374
$4A$	$+$	$2C$	$=$	374

admission for 2 adults	and	admission for 3 children	is	$285
$2A$	$+$	$3C$	$=$	285

SOLVE and INTERPRET.

Multiply the 2nd equation by -2.

$$\begin{cases} 4A + 2C = 374 \\ 2A + 3C = 285 \end{cases} \rightarrow \begin{cases} 4A + 2C = 374 \\ -4A - 6C = -570 \end{cases}$$
$$-4C = -196$$
$$C = \$49$$

Now, you find A:

Your turn:

4. Ann Marie Jones has been pricing Amtrak train fares for a group trip to New York. Three adults and four children must pay $159. Two adults and three children must pay $112. Find the price of an adult's ticket, and find the price of a child's ticket.

	Answer	Text Ref	Video Ref		Answer	Text Ref	Video Ref
1	29 and 8	Ex 1, p. 875-876		3	Child's ticket: $49 Adult's ticket: $69	Ex 2, p. 876-877	
2	33 and 50		Sec 11.4, 1/5	4	Child's ticket: $18 Adult's ticket: $29		Sec 11.4, 2/5

☐ **Next, insert your homework.** Make sure you attempt all exercises asked of you and show all work, as in the exercises above. Check your answers if possible. Clearly mark any exercises you were unable to correctly complete so that you may ask questions later. DO NOT ERASE YOUR INCORRECT WORK. THIS IS HOW WE UNDERSTAND AND EXPLAIN TO YOU YOUR ERRORS.

Preparing for the Chapter 11 Test

Start preparing for your Chapter 11 Test as soon as possible. Pay careful attention to any instructor discussion about this test, especially discussion on what sections you will be responsible for, etc.

☐ Work the Chapter 11 Vocabulary Check on page 887.

☐ Read both columns (Definitions and Concepts, and Examples) of the Chapter 11 Highlights starting on page 887.

☐ Read your Class Notes/Examples for each section covered on your Chapter 11 Test. Look for any unresolved questions you may have.

☐ Complete as many of the Chapter 11 Review exercises as possible (starting on page 890). Remember, the odd answers are in the back of your text.

☐ **Most important:** Place yourself in "test" conditions (see below) and work the Chapter 11 Test (page 893-894) as a practice test the day before your actual test. To honestly assess how you are doing, try the following:

- Work on a few blank sheets of paper.
- Give yourself the same amount of time you will be given for your actual test.
- Complete this Chapter 11 Practice Test without using your notes or your text.
- If you have any time left after completing this practice test, check your work and try to find any errors on your own.
- Once done, use the back of your book to check ALL answers.
- Try to correct any errors on your own.
- Use the Chapter Test Prep Video (CTPV) to correct any errors you were unable to correct on your own. You can find these videos in the Interactive DVD Lecture Series, in MyMathLab, and on YouTube. Search MartinGayDevMath and click "Channels."

I wish you the best of luck….Elayn Martin-Gay

Section 12.1 Exponents

Before Class:

☐   Read the objectives on page 898.

☐   Read the **Helpful Hint** boxes on pages 898, 900, and 901.

☐   Complete the exercises:

1. In $2^4 = 16$, the 2 is called the _____ and the 4 is called the _____ .

2. Complete the list of common perfect squares.

$1^2 = $ _____ $5^2 = $ _____ $9^2 = $ _____

$2^2 = $ _____ $6^2 = $ _____ $10^2 = $ _____

$3^2 = $ _____ $7^2 = $ _____ $11^2 = $ _____

$4^2 = $ _____ $8^2 = $ _____ $12^2 = $ _____

3. Evaluate the expression: $(-2)^3 = (-2)(-2)(-2) = $ _____

During Class:

☐   **Write your class notes.** Neatly write down **all** examples shown as well as key terms or phrases with definitions. If not applicable or if you were absent, watch the Lecture Series (DVD) for this section and do the same (write down the examples shown as well as key terms or phrases). Insert more paper as needed.

Class Notes/Examples	**Your Notes**

Answers: **1)** base, exponent   **2)** $1, 4, 9, 16, 25, 36, 49, 64, 81, 100, 121, 144$ **3)** -8

Section 12.1 Exponents

Class Notes (continued)	**Your Notes**

(Insert additional paper as needed.)

Practice:

☐   Complete the Vocabulary and Readiness Check on page 906.

☐   Next, complete any incomplete exercises below. Check and correct your work using the answers and references at the end of this section.

Review this example:

1. Evaluate each expression for the given value of x.

    a. $2x^3$ when x is 5

    b. $\dfrac{9}{x^2}$ when x is -3

a. $2x^3 = 2\cdot 5^3 = 2\cdot(5\cdot5\cdot5) = 2\cdot125 = 250$

b. $\dfrac{9}{x^2} = \dfrac{9}{(-3)^2} = \dfrac{9}{(-3)(-3)} = \dfrac{9}{9} = 1$

Your turn:

2. Evaluate the expression for the given value of z.

    $\dfrac{2z^4}{5}$ when z is -2

Review this example:

3. Use the product rule to simplify each expression.

    a. $(2x^2)(-3x^5)$

    b. $(-a^7b^4)(3ab^9)$

a. $(2x^2)(-3x^5)$ Group like bases.

   $= (2\cdot x^2)\cdot(-3\cdot x^5)$

   $= (2\cdot-3)\cdot(x^2\cdot x^5)$ Add exponents and simplify.

   $= -6x^7$

b. $(-a^7b^4)(3ab^9)$ Group like bases.

   $= (-1\cdot3)(a^7\cdot a^1)(b^9\cdot b^4)$ Write a as a^1. Add exponents and simplify.

   $= -3a^8b^{13}$

Your turn:

4. Use the product rule to simplify each expression.

    a. $(5y^4)(3y)$

    b. $(x^9y)(x^{10}y^5)$

307

Section 12.1 Exponents

Complete these examples:

5. Simplify each expression.

a. $(2a)^3$ b. $\left(\dfrac{2x^4}{3y^5}\right)^4$ c. $\dfrac{2x^5 y^2}{xy}$

Use power of product rule.

a. $(2a)^3 = 2^3 \cdot a^3 = \boxed{\underline{} a^3}$

Use power of quotient rule.

b. $\left(\dfrac{2x^4}{3y^5}\right)^4 = \dfrac{2^4 \cdot (x^4)^4}{3^4 \cdot (y^5)^4} = \boxed{\dfrac{16x^{16}}{81y^{20}}, \, y \ne 0}$

Use the quotient rule.

c. $\dfrac{2x^5 y^2}{xy} = 2 \cdot \dfrac{x^5}{x^1} \cdot \dfrac{y^2}{y^1} = 2 \cdot \left(x^{5-1}\right) \cdot \left(y^{2-1}\right) = \boxed{\underline{}}$

Your turn:

6. Simplify each expression.

a. $(x^2 y^3)^5$

b. $\left(\dfrac{-2xz}{y^5}\right)^2$

c. $\dfrac{9a^4 b^7}{27ab^2}$

Review this example:

7. Simplify the expression $(-4)^0$.

A base raised to the zero power is 1. (base $\ne 0$)

$(-4)^0 = 1$

Your turn:

8. Simplify the expression $(2x)^0$.

	Answer	Text Ref	Video Ref	Answer	Text Ref	Video Ref
1	a. 250 b. 1	Ex 7a, b, p. 899		**5**	a. $8a^3$; b. $\dfrac{16x^{16}}{81y^{20}}$ $y \ne 0$; c. $2x^4 y$	Ex 19, 23, 27, p. 902–904	
2	$\dfrac{32}{5}$		Sec 12.1, 8/30	**6**	a. $x^{10} y^{15}$; b. $\dfrac{4x^2 z^2}{y^{10}}$ c. $\dfrac{a^3 b^5}{3}$		Sec 12.1, 19, 20, 23/30
3	a. $-6x^7$ b. $-3a^8 b^{13}$	Ex 13, 15, p. 900		**7**	1	Ex 30, p. 904	
4	a. $15y^5$ b. $x^{19} y^6$		Sec 12.1, 12–13/30	**8**	1		Sec 12.1, 25/30

☐ **Next, insert your homework.** Make sure you attempt all exercises asked of you and show all work, as in the exercises above. Check your answers if possible. Clearly mark any exercises you were unable to correctly complete so that you may ask questions later. DO NOT ERASE YOUR INCORRECT WORK. THIS IS HOW WE UNDERSTAND AND EXPLAIN TO YOU YOUR ERRORS.

Section 12.2 Negative Exponents and Scientific Notation

Before Class:

☐   Read the objectives on page 910.

☐   Read the **Helpful Hint** boxes on page 910 and 911.

☐   Complete the exercises:

     1.   Using the product rule, $x^4 x^5$ simplifies to _____ .

     2.   Using the product rule, $x^4 x^5 x$ simplifies to _____ .

     3.   Using the quotient rule, $\dfrac{x^8}{x^2}$ simplifies to _____ .

     4.   Using the power rule, $\left(x^4\right)^5$ simplifies to _____ .

     5.   Using the zero exponent rule, 3^0 simplifies to _____ .

During Class:

☐   **Write your class notes.** Neatly write down **all** examples shown as well as key terms or phrases with definitions. If not applicable or if you were absent, watch the Lecture Series (DVD) for this section and do the same (write down the examples shown as well as key terms or phrases). Insert more paper as needed.

Class Notes/Examples	Your Notes

Answers: **1)** x^9 **2)** x^{10} **3)** x^6 **4)** x^{20} **5)** 1

Section 12.2 Negative Exponents and Scientific Notation

Class Notes (continued)	Your Notes

(Insert additional paper as needed.)

Section 12.2 Negative Exponents and Scientific Notation

Practice:

☐   Complete the Vocabulary and Readiness Check on page 916.

☐   Next, complete any incomplete exercises below. Check and correct your work using the answers and references at the end of this section.

Review this example:

1. Simplify each expression. Write each result using positive exponents only.

a. $\dfrac{y}{y^{-2}}$

Use the quotient rule.

↓

$\dfrac{y}{y^{-2}} = \dfrac{y^1}{y^{-2}} = y^{1-(-2)} = y^{1+2} = \boxed{y^3}$

b. $\dfrac{x^{-5}}{x^7}$

Use the quotient rule and negative exponent rule.

↓   ↓

$\dfrac{x^{-5}}{x^7} = x^{-5-7} = x^{-12} = \boxed{\dfrac{1}{x^{12}}}$

Your turn:

2. Simplify the expression. Write your result using positive exponents only.

$\dfrac{r}{r^{-3}r^{-2}}$

Complete this example:

3. Simplify: $\dfrac{(2xy)^{-3}}{\left(x^2 y^3\right)^2}$

Raise each factor in the numerator to the −3 power and raise each factor in the denominator to the second power. Then, use the quotient rule.

$\dfrac{(2xy)^{-3}}{\left(x^2 y^3\right)^2} = \dfrac{2^{-3} x^{-3} y^{-3}}{x^4 y^6} = 2^{-3} x^{-3-4} y^{-3-6} = 2^{-3} x^{-7} y^{-9}$

Now, use the negative exponent rule:

$2^{-3} x^{-7} y^{-9} = \bigcirc$

Your turn:

4. Simplify the expression. Write your result using positive exponents only.

$\dfrac{\left(-2xy^{-3}\right)^{-3}}{\left(xy^{-1}\right)^{-1}}$

311

Section 12.2 Negative Exponents and Scientific Notation

Review this example:

5. Write $367{,}000{,}000$ in scientific notation.

Move the decimal point 8 places so that it is between the 3 and the 6. The new number 3.67 is now between 1 and 10.

original number
↓
$$367{,}000{,}000 \;=\; 3.67 \times 10^{?}$$

The original number is 10 or greater, so the exponent of 10 will be a positive 8.

scientific notation
↓
$$367{,}000{,}000 \;=\; \boxed{3.67 \times 10^{8}}$$

Your turn:

6. Write in scientific notation.

a. $78{,}000$

b. 0.00000167

Review this example:

9. Write 7.358×10^{-3} in standard notation.

Move the decimal point <u>left</u> 3 places.

$$7.358 \times 10^{-3} = \boxed{0.007358}$$

Your turn:

10. Write the number in standard notation.

$$2.032 \times 10^{4}$$

	Answer	Text Ref	Video Ref		Answer	Text Ref	Video Ref
1	a. y^3 b. $\dfrac{1}{x^{12}}$	Ex 6, 8, p. 911		**5**	3.67×10^8	Ex 17a, p. 913	
2	r^6		Sec 12.2, 7/13	**6**	a. 7.8×10^4 b. 1.67×10^{-6}		Sec 12.2, 9–10/13
3	$\dfrac{1}{8x^7 y^9}$	Ex 16, p. 912		**7**	0.007358	Ex 18b, p. 914	
4	$-\dfrac{y^8}{8x^2}$		Sec 12.2, 8/7	**8**	$20{,}320$		Sec 12.2, 12/13

☐   **Next, insert your homework.** Make sure you attempt all exercises asked of you and show all work, as in the exercises above. Check your answers if possible. Clearly mark any exercises you were unable to correctly complete so that you may ask questions later. DO NOT ERASE YOUR INCORRECT WORK. THIS IS HOW WE UNDERSTAND AND EXPLAIN TO YOU YOUR ERRORS.

Section 12.3 Introduction to Polynomials

Before Class:

☐   Read the objectives on page 920.

☐   Read the **Helpful Hint** boxes on pages 924 and 925.

☐   Complete the exercises:

1. Using the quotient rule and negative exponent rule, $\dfrac{x^{-4}}{x^7}$ simplifies to _____ .

2. Using the negative exponent rule, $2^{-2} x^{-2} y^{-3}$ simplifies to _____ .

3. Using the zero exponent rule, $2^0 - 3^0$ simplifies to _____ .

4. Write 9.7×10^{-2} in standard notation: _____

    Write 1.3×10^1 in standard notation: _____

5. Which is larger? 8.6×10^5 or 4.4×10^7

During Class:

☐   **Write your class notes.** Neatly write down **all** examples shown as well as key terms or phrases with definitions. If not applicable or if you were absent, watch the Lecture Series (DVD) for this section and do the same (write down the examples shown as well as key terms or phrases). Insert more paper as needed.

Class Notes/Examples	**Your Notes**

Answers: **1)** $\dfrac{1}{x^{11}}$ **2)** $\dfrac{1}{4x^2 y^3}$ **3)** 0   **4)** .097; 13   **5)** 4.4×10^7

Section 12. 3 Introduction to Polynomials

Class Notes (continued)	Your Notes

(Insert additional paper as needed.)

Practice:

☐   Complete the Vocabulary and Readiness Check on page 726.

☐   Next, complete any incomplete exercises below. Check and correct your work using the answers and references at the end of this section.

Review this example:

1. Complete the table for the expression $7x^5 - 8x^4 + x^2 - 3x + 5$.

Term	Coefficient
x^2	
	-8
$-3x$	
	7
5	

The completed table is shown below.

Term	Coefficient
x^2	1
$-8x^4$	-8
$-3x$	-3
$7x^5$	7
5	5

Your turn:

2. Complete the table for the expression $x^2 - 3x + 5$.

Term	Coefficient
x^2	
	-3
5	

Complete this example:

3. Find the degree of the polynomial $7x + 3x^3 + 2x^2 - 1$. Tell whether the polynomial is a monomial, binomial, trinomial or none of these.

The degree of a polynomial is the greatest degree of any term of the polynomial:

The term $7x$ is $7x^1$ has degree 1.
The term $3x^3$ has degree 3.
The term $2x^2$ has degree 2.
The term -1 is $-1x^0$ has degree 0.

The degree is ⬭ and the polynomial is neither a monomial, binomial, or trinomial.

Your turn:

4. Find the degree of the polynomial $9m^3 - 5m^2 + 4m - 8$. Tell whether the polynomial is a monomial, binomial, trinomial or none of these.

Section 12. 3 Introduction to Polynomials

Complete this example:

5. Evaluate the polynomial $3x^2 - 2x + 1$ when $x = -2$.

Replace x with -2.

$3x^2 - 2x + 1 = 3(-2)^2 - 2(-2) + 1$
$\qquad\qquad = 3(4) + 4 + 1$
$\qquad\qquad = \underline{\qquad\qquad} = \underline{\quad}$

Your turn:

6. Evaluate the polynomial $x^2 - 5x - 2$ when (a) $x = 0$ and (b) $x = -1$.

a.

b.

Review this example:

7. Simplify the given polynomial by combining like terms: $11x^2 + 5 + 2x^2 - 7$

$11x^2 + 5 + 2x^2 - 7 = 11x^2 + 2x^2 + 5 - 7$
$\qquad\qquad\qquad\qquad = 13x^2 - 2$

Your turn:

8. Simplify the given polynomial by combining like terms:

$3ab - 4a + 6ab - 7a$

Review this example:

9. Write the given polynomial in descending powers of the variable with no missing powers: $2x + x^4$

$2x + x^4 = x^4 + 2x$ Write in descending powers.
$\qquad\quad = x^4 + 0x^3 + 0x^2 + 2x + 0x^0$

Insert missing terms of $0x^3, 0x^2$, and $0x^0$ (or 0).

Your turn:

10. Write the given polynomial in descending powers of the variable with no missing powers:

$5y^3 + 2y - 10$

	Answer	Text Ref	Video Ref		Answer	Text Ref	Video Ref
1	pictured above	Ex 1, p. 920		6	a. -2 b. 4		Sec 12.3, 3–4/7
2	$1; -3x; 5$		Sec 12.3, 1/7	7	$13x^2 - 2$	Ex 7, p. 923	
3	3 ; none of these	Ex 3c, p. 922		8	$9ab - 11a$		Sec 12.3, 6/7
4	3 ; none of these		Sec 12.3, 2/7	9	$x^4 + 0x^3 + 0x^2 + 2x + 0x^0$	Ex 15c, p. 925	
5	17	Ex 4b, p. 922		10	$5y^3 + 0y^2 + 2y - 10$		Sec 12.3, 7/6

☐   **Next, insert your homework.** Make sure you attempt all exercises asked of you and show all work, as in the exercises above. Check your answers if possible. Clearly mark any exercises you were unable to correctly complete so that you may ask questions later. DO NOT ERASE YOUR INCORRECT WORK. THIS IS HOW WE UNDERSTAND AND EXPLAIN TO YOU YOUR ERRORS.

Section 12.4 Adding and Subtracting Polynomials

Before Class:

☐   Read the objectives on page 931.

☐   Read the **Helpful Hint** box on page 932.

☐   Complete the exercises:

1.   Combine like terms: $3x^2 - 7x^2 = $ _____

2.   Combine like terms: $-3x^2 - 7x^2 - 4x - 6x = $ _____

3.   Combine like terms: $-3x^2 + 7x^2 + 4x - 6x = $ _____

4.   Combine like terms: $3x + 4xy - 4xy + 6x - 3 + 5 = $ _____

5.   Combine like terms: $2x^2 + 4x - 3x^2 + 6 - 3x + 5 = $ _____

During Class:

☐   **Write your class notes.** Neatly write down **all** examples shown as well as key terms or phrases with definitions. If not applicable or if you were absent, watch the Lecture Series (DVD) for this section and do the same (write down the examples shown as well as key terms or phrases). Insert more paper as needed.

Class Notes/Examples	**Your Notes**

Answers: **1)** $-4x^2$ **2)** $-10x^2 - 10x$ **3)** $4x^2 - 2x$ **4)** $9x + 2$ **5)** $-x^2 + x + 11$

Class Notes (continued)

Your Notes

(Insert additional paper as needed.)

Section 12.4 Adding and Subtracting Polynomials

Practice:

☐   Complete the Vocabulary and Readiness Check on page 934.

☐   Next, complete any incomplete exercises below. Check and correct your work using the answers and references at the end of this section.

Review this example:

1. Add: $\left(-2x^2 + 5x - 1\right) + \left(-2x^2 + x + 3\right)$

Remove parentheses.

$\left(-2x^2 + 5x - 1\right) + \left(-2x^2 + x + 3\right)$
$= \ -2x^2 + 5x - 1 - 2x^2 + x + 3$

Group like terms together. Then, combine like terms.

$= \ -2x^2 - 2x^2 + 5x + x - 1 + 3$

$= \boxed{-4x^2 + 6x + 2}$

Your turn:

2. Add: $\left(-3y^2 - 4y\right) + \left(2y^2 + y - 1\right)$

Complete this example:

3. Subtract: $\left(2x^3 + 8x^2 - 6x\right) - \left(2x^3 - x^2 + 1\right)$

Remove parentheses and change the sign of each term in the second polynomial.

$\left(2x^3 + 8x^2 - 6x\right) - \left(2x^3 - x^2 + 1\right)$
$= \ 2x^3 + 8x^2 - 6x - 2x^3 + x^2 - 1$

Now, combine like terms:

$=$

Your turn:

4. Subtract: $\left(5x + 8\right) - \left(-2x^2 - 6x + 8\right)$

319

Section 12.4 Adding and Subtracting Polynomials

Review this example:

5. Subtract $(5y^2 + 2y - 6)$ from

$(-3y^2 - 2y + 11)$ using a vertical format.

Change the sign of each term in the polynomial being subtracted.

Add like terms.

$-3y^2 - 2y + 11$ $-3y^2 - 2y + 11$

$-(5y^2 + 2y - 6)$ $-5y^2 - 2y + 6$

$\boxed{-8y^2 - 4y + 17}$

Your turn:

6. Subtract $(5x + 7)$ from

$(7x^2 + 3x + 9)$ using a vertical format.

Review this example:

7. Add:

$(3x^2 - 6xy + 5y^2) + (-2x^2 + 8xy - y^2)$

Remove the parentheses and combine like terms.

$(3x^2 - 6xy + 5y^2) + (-2x^2 + 8xy - y^2)$

$= 3x^2 - 6xy + 5y^2 - 2x^2 + 8xy - y^2$

$= \boxed{x^2 + 2xy + 4y^2}$

Your turn:

8. Add:

$(x^2 + 2xy - y^2) + (5x^2 - 4xy + 20y^2)$

	Answer	Text Ref	Video Ref		Answer	Text Ref	Video Ref
1	$-4x^2 + 6x + 2$	Ex 2, p. 931		5	$-8y^2 - 4y + 17$	Ex 6, p. 932	
2	$-y^2 - 3y - 1$		Sec 12.4, 1/5	6	$7x^2 - 2x + 2$		Sec 12.4, 3/5
3	$9x^2 - 6x - 1$	Ex 5, p. 932		7	$x^2 + 2xy + 4y^2$	Ex 8, p. 933	
4	$2x^2 + 11x$		Sec 12.4, 2/5	8	$6x^2 - 2xy + 19y^2$		Sec 12.4, 5/5

☐ **Next, insert your homework.** Make sure you attempt all exercises asked of you and show all work, as in the exercises above. Check your answers if possible. Clearly mark any exercises you were unable to correctly complete so that you may ask questions later. DO NOT ERASE YOUR INCORRECT WORK. THIS IS HOW WE UNDERSTAND AND EXPLAIN TO YOU YOUR ERRORS.

Before Class:

☐   Read the objectives on page 938.

☐   Complete the exercises:

1.   Simplify each expression by performing the indicated operation.

    a. $z + 3z =$ _____   b. $z \cdot 3z =$ _____

    c. $-z - 3z =$ _____   d. $(-z)(-3z) =$ _____

2.   Simplify each expression by performing the indicated operation.

    a. $m \cdot m \cdot m =$ _____   b. $m + m + m =$ _____

    c. $(-m)(-m)(-m) =$ _____   d. $-m - m - m =$ _____

During Class:

☐   **Write your class notes.** Neatly write down **all** examples shown as well as key terms or phrases with definitions. If not applicable or if you were absent, watch the Lecture Series (DVD) for this section and do the same (write down the examples shown as well as key terms or phrases). Insert more paper as needed.

Class Notes/Examples	**Your Notes**

Answers: **1)** a. $4z$ b. $3z^2$ c. $-4z$ d. $3z^2$ **2)** a. m^3 b. $3m$ c. $-m^3$ d. $-3m$

Section 12.5 Multiplying Polynomials

Class Notes (continued)	**Your Notes**

(Insert additional paper as needed.)

Practice:

☐   Complete the Vocabulary and Readiness Check on page 941.

☐   Next, complete any incomplete exercises below. Check and correct your work using the answers and references at the end of this section.

Review this example:
1. Multiply.

a. $-7x^2 \cdot 2x^5$

b. $(-12x^5)(-x)$

Use commutative and associative properties.

a. $-7x^2 \cdot 2x^5 = (-7 \cdot 2)(x^2 \cdot x^5)$
$= \boxed{-14x^7}$

b. $(-12x^5)(-x) = (-12x^5)(-1x)$
$\qquad\qquad = (-12)(-1)(x^5 \cdot x)$
$\qquad\qquad \boxed{= 12x^6}$

Your turn:
2. Multiply.

a. $6x \cdot 3x^2$

b. $\left(-\dfrac{1}{3}y^2\right)\left(\dfrac{2}{5}y\right)$

Review this example:
3. Multiply.

a. $5x(2x^3 + 6)$

b. $-3x^2(5x^2 + 6x - 1)$

Apply the distributive property.

a. $5x(2x^3 + 6) = 5x(2x^3) + 5x(6)$ Multiply.
$\qquad\qquad\quad \boxed{= 10x^4 + 30x}$

b. $-3x^2(5x^2 + 6x - 1)$
$= (-3x^2)(5x^2) + (-3x^2)(6x) + (-3x^2)(-1)$
$\boxed{= -15x^4 - 18x^3 + 3x^2}$

Your turn:
4. Multiply.

a. $3x(2x + 5)$

b. $-y(4x^3 - 7x^2y + xy^2 + 3y^3)$

323

Section 12.5 Multiplying Polynomials

Complete this example:

5. Multiply: $(3x+2)(2x-5)$

Apply the distributive property twice.

$= 3x(2x-5) + 2(2x-5)$ Distribute.

$= 3x(2x) + 3x(-5) + 2(2x) + 2(-5)$ Distribute.

$= 6x^2 - 15x + 4x - 10$ Multiply.

$=$ ⬭   Combine like terms.

Your turn:

6. Multiply.

  a. $(a+7)(a-2)$

Apply the definition of an exponent.

  b. $(7xy - y)^2 = ($_____$)($_____$)$

Review this example:

7. Multiply vertically: $(2y^2 + 5)(y^2 - 3y + 4)$

Step 1. Multiply $y^2 - 3y + 4$ by 5.

Step 2. Multiply $y^2 - 3y + 4$ by $2y^2$.

Step 3. Combine like terms.

$$
\begin{array}{r}
y^2 - 3y + 4 \\
2y^2 + 5 \\
\hline
5y^2 - 15y + 20 \\
2y^4 - 6y^3 + 8y^2 \\
\hline
2y^4 - 6y^3 + 13y^2 - 15y + 20
\end{array}
$$

  Step 1

  Step 2

  Step 3

Your turn:

8. Multiply vertically.

  $(x+3)(2x^2 + 4x - 1)$

	Answer	Text Ref	Video Ref		Answer	Text Ref	Video Ref
1	a. $-14x^7$ b. $12x^6$	Ex 2, 3, p. 938		5	$6x^2 - 11x - 10$	Ex 7b, p. 939	
2	a. $18x^3$ b. $-\dfrac{2}{15}y^3$		Sec 12.5, 1–2/8	6	a. $a^2 + 5a - 14$ b. $49x^2y^2 - 14xy^2 + y^2$		Sec 12.5, 5–6/8
3	a. $10x^4 + 30x$ b. $-15x^4 - 18x^3 + 3x^2$	Ex 5, 6, p. 938		7	$2y^4 - 6y^3 + 13y^2 - 15y + 20$	Ex 10, p. 940	
4	a. $6x^2 + 15x$ b. $-4x^3y + 7x^2y^2$ $-xy^3 - 3y^4$		Sec 12.5, 3–4/8	8	$2x^3 + 10x^2 + 11x - 3$		Sec 12.5, 8/8

☐   **Next, insert your homework.** Make sure you attempt all exercises asked of you and show all work, as in the exercises above. Check your answers if possible. Clearly mark any exercises you were unable to correctly complete so that you may ask questions later. DO NOT ERASE YOUR INCORRECT WORK. THIS IS HOW WE UNDERSTAND AND EXPLAIN TO YOU YOUR ERRORS.

Section 12.6 Special Products

Before Class:

☐ Read the objectives on page 945.

☐ Read the **Helpful Hint** boxes on pages 945, 947, and 948.

☐ Complete the exercises:

1. Multiply: $3(-7x) =$ _____

2. Multiply: $3x(-7x) =$ _____

3. Multiply: $3x(-7x - 6) =$ _____

4. Combine like terms: $16x^2 - 36x - 36x + 81 =$ _____

5. Combine like terms: $16x^2 + 36x - 36x - 81 =$ _____

During Class:

☐ **Write your class notes.** Neatly write down **all** examples shown as well as key terms or phrases with definitions. If not applicable or if you were absent, watch the Lecture Series (DVD) for this section and do the same (write down the examples shown as well as key terms or phrases). Insert more paper as needed.

Class Notes/Examples	**Your Notes**

Answers: **1)** $-21x$　　**2)** $-21x^2$　　**3)** $-21x^2 - 18x$　　**4)** $16x^2 - 72x + 81$　　**5)** $16x^2 - 81$

325

Section 12.6 Special Products

Class Notes (continued)

Your Notes

(Insert additional paper as needed.)

Practice:

☐   Complete the Vocabulary and Readiness Check on page 949.

☐   Next, complete any incomplete exercises below. Check and correct your work using the answers and references at the end of this section.

Review this example:	**Your turn:**
1. Multiply: $(5x-7)(x-2)$	**2.** Multiply: $(x+3)(x+4)$

$$\underset{\overset{\uparrow}{\underset{\text{O}}{\underset{\uparrow}{\text{I}}}}}{\overset{\text{F}\;\;\;\;\text{L}}{(5x-7)(x-2)}}$$

$$\quad\;\; \textbf{F} \quad\; \textbf{O} \quad\;\; \textbf{I} \quad\;\; \textbf{L}$$
$$= 5x(x) + 5x(-2) + (-7)x + (-7)(-2)$$
$$= 5x^2 - 10x - 7x + 14$$
$$= \boxed{5x^2 - 17x + 14}$$

Multiply. Then, combine like terms.

Complete this example:	**Your turn:**
3. Multiply: $(3y+1)^2$	**4.** Multiply: $(2x-1)^2$

Since $(3y+1)^2 = (3y+1)(3y+1)$ we can use the **FOIL** method.

$$(3y+1)^2 = (3y+1)(3y+1)$$

$$\quad\;\; \textbf{F} \quad\;\; \textbf{O} \quad\; \textbf{I} \quad\;\; \textbf{L}$$
$$= 3y(3y) + 3y(1) + 1(3y) + 1(1)$$
$$= 9y^2 + 3y + 3y + 1 \qquad \text{Multiply.}$$

$$= \underset{}{\bigcirc}$$

Now, you combine like terms.

327

Section 12.6 Special Products

Complete this example:

5. Multiply: $(6t+7)(6t-7)$

Method 1: FOIL Method

$$\overset{\textbf{F}\qquad\textbf{O}\qquad\textbf{I}\qquad\textbf{L}}{(6t+7)(6t-7)=}\ 6t(6t)+6t(-7)+7(6t)+7(-7)$$

$$= 36t^2 - 42t + 42t - 49 \qquad \text{Multiply.}$$

$= \bigcirc$ Now, you combine like terms.

Method 2: $(a+b)(a-b)=a^2-b^2$

$$(6t+7)(6t-7) = (6t)^2 - (7)^2$$

$= \bigcirc$ Simplify by squaring each term.

Your turn:

6. Multiply: $(4x+5)(4x-5)$

Method 1: FOIL Method

Method 2: $(a+b)(a-b)=a^2-b^2$

	Answer	Text Ref	Video Ref		Answer	Text Ref	Video Ref
1	$5x^2-17x+14$	Ex 2, p. 945		4	$4x^2-4x+1$		Sec 12.6, 4/8
2	$x^2+7x+12$		Sec 12.6, 1/8	5	$36t^2-49$	Ex 10, p. 948	
3	$9y^2+6y+1$	Ex 4, p.946		6	$16x^2-25$		Sec 12.6, 7/8

☐ **Next, insert your homework.** Make sure you attempt all exercises asked of you and show all work, as in the exercises above. Check your answers if possible. Clearly mark any exercises you were unable to correctly complete so that you may ask questions later. DO NOT ERASE YOUR INCORRECT WORK. THIS IS HOW WE UNDERSTAND AND EXPLAIN TO YOU YOUR ERRORS.

Section 12.7 Dividing Polynomials

Before Class:

☐   Read the objectives on page 954.

☐   Read the **Helpful Hint** box on page 955.

☐   Complete the exercises:

    1. Simplify: $\dfrac{a+7}{7} = \dfrac{a}{7} + \dfrac{7}{7} =$ _____

    2. Simplify: $\dfrac{5x+15}{5} = \dfrac{5x}{5} + \dfrac{15}{5} =$ _____

    3. Simplify: $\dfrac{x+5}{5} = - \ + \ - \ =$ _____

During Class:

☐   **Write your class notes.** Neatly write down **all** examples shown as well as key terms or phrases with definitions. If not applicable or if you were absent, watch the Lecture Series (DVD) for this section and do the same (write down the examples shown as well as key terms or phrases). Insert more paper as needed.

Class Notes/Examples	**Your Notes**

Answers: **1)** $\dfrac{a}{7}+1$ **2)** $x+3$ **3)** $\dfrac{x}{5}+\dfrac{5}{5}=\dfrac{x}{5}+1$

Section 12.7 Dividing Polynomials

Class Notes (continued)	Your Notes

(Insert additional paper as needed.)

Section 12.7 Dividing Polynomials

Practice:

☐ Complete the Vocabulary and Readiness Check on page 958.

☐ Next, complete any incomplete exercises below. Check and correct your work using the answers and references at the end of this section.

Review this example:

1. Divide: $(6m^2 + 2m) \div 2m$

Write the quotient in fraction form.
Then divide each term of the polynomial by the monomial.

$$\frac{6m^2 + 2m}{2m} = \frac{6m^2}{2m} + \frac{2m}{2m}$$

$$= 3m + 1 \qquad \text{Simplify.}$$

Your turn:

2. Divide: $\dfrac{12x^4 + 3x^2}{x}$

Review this example:

3. Divide: $\dfrac{9x^5 - 12x^2 + 3x}{3x^2}$

Divide each term by $3x^2$.

$$\frac{9x^5 - 12x^2 + 3x}{3x^2} = \frac{9x^5}{3x^2} - \frac{12x^2}{3x^2} + \frac{3x}{3x^2}$$

$$= 3x^3 - 4 + \frac{1}{x} \qquad \text{Simplify.}$$

Check: To check, we multiply.

$$3x^2 \left(3x^3 - 4 + \frac{1}{x} \right) = 3x^2 (3x^3) - 3x^2 (4) + 3x^2 \left(\frac{1}{x} \right)$$

$$= 9x^5 - 12x^2 + 3x$$

Your turn:

4. Divide: $\dfrac{-9x^5 + 3x^4 - 12}{3x^3}$

331

Section 12.7 Dividing Polynomials

Review this example:

5. Divide: $\dfrac{x^2+7x+12}{x+3}$

$$x+3\overline{\smash{\big)}\,x^2+7x+12} \quad \begin{array}{l} x \end{array}$$
$$\underline{x^2+3x}\quad\downarrow$$
$$4x+12$$

How many times does x divide x^2?

$\dfrac{x^2}{x}=x$

Multiply: $x(x+3)$

Subtract. Bring down the next term.

Repeat the process.

$$x+3\overline{\smash{\big)}\,x^2+7x+12}\quad\begin{array}{l}x\;+\;4\end{array}$$
$$\underline{x^2+3x}\quad\downarrow$$
$$4x+12$$
$$\rightarrow\;\rightarrow\;\underline{4x\!\!\not/\,12}$$
$$0$$

How many times does x divide $4x$?

$\dfrac{4x}{x}=4$

Multiply: $4(x+3)$

Subtract. The remainder is 0.

The quotient is $\boxed{x+4}$

Your turn:

6. Divide: $\dfrac{x^2+4x^4+3}{x+3}$

Review this example:

7. Divide x^3-8 by $x-2$.

$$x-2\overline{\smash{\big)}\,x^3+0x^2+0x-8}\quad\begin{array}{l}x^2+2x+4\end{array}$$
$$-\;\underline{x^3\!\!\not/\,2x^2}$$
$$2x^2+0x$$
$$-\;\underline{2x^2\!\!\not/\,4x}$$
$$4x-8$$
$$-\;\underline{4x\!\!\not/\,8}$$
$$0$$

Notice that the x^2 and x terms are missing. Insert $0x^2$ and $0x$ into the polynomial.

The quotient is $\boxed{x^2+2x+4\,.}$

Your turn:

8. Divide: $\dfrac{x^3-27}{x-3}$

	Answer	Text Ref	Video Ref		Answer	Text Ref	Video Ref
1	$3m+1$	Ex 1, p. 954		**5**	$x+4$	Ex 4, p. 955–956	
2	$12x^3+3x$		Sec 12.7, 1/5	**6**	$x+1$		Sec 12.7, 3/5
3	$3x^3-4+\dfrac{1}{x}$	Ex 2, p. 954		**7**	x^2+2x+4	Ex 8, p. 957	
4	$-3x^2+x-\dfrac{4}{x^3}$		Sec 12.7, 2/5	**8**	x^2+3x+9		Sec 12.7, 5/5

☐ **Next, insert your homework.** Make sure you attempt all exercises asked of you and show all work, as in the exercises above. Check your answers if possible. Clearly mark any exercises you were unable to correctly complete so that you may ask questions later. DO NOT ERASE YOUR INCORRECT WORK. THIS IS HOW WE UNDERSTAND AND EXPLAIN TO YOU YOUR ERRORS.

Preparing for the Chapter 12 Test

Start preparing for your Chapter 12 Test as soon as possible. Pay careful attention to any instructor discussion about this test, especially discussion on what sections you will be responsible for, etc.

☐ Work the Chapter 12 Vocabulary Check on page 962.

☐ Read both columns (Definitions and Concepts, and Examples) of the Chapter 12 Highlights starting on page 962.

☐ Read your Class Notes/Examples for each section covered on your Chapter 12 Test. Look for any unresolved questions you may have.

☐ Complete as many of the Chapter 12 Review exercises as possible (page 965). Remember, the odd answers are in the back of your text.

☐ **Most important:** Place yourself in "test" conditions (see below) and work the Chapter 12 Test (page 970 - 971) as a practice test the day before your actual test. To honestly assess how you are doing, try the following:

- Work on a few blank sheets of paper.
- Give yourself the same amount of time you will be given for your actual test.
- Complete this Chapter 12 Practice Test without using your notes or your text.
- If you have any time left after completing this practice test, check your work and try to find any errors on your own.
- Once done, use the back of your book to check ALL answers.
- Try to correct any errors on your own.
- Use the Chapter Test Prep Video (CTPV) to correct any errors you were unable to correct on your own. You can find these videos in the Interactive DVD Lecture Series, in MyMathLab, and on YouTube. Search Martin-Gay Prealgebra & Introductory Algebra and click "Channels."

I wish you the best of luck….Elayn Martin-Gay

Section 13.1 The Greatest Common Factor

Before Class:

☐   Read the objectives on page 976.

☐   Read the **Helpful Hint** boxes on pages 977, 978, 979, 980 and 981.

☐   Complete the exercises:

1.   The GCF of a list of terms contains the _____ exponent on each common variable.

2.   List all of the factors of 45: _____ .

       List all of the factors of 75: _____ .

       What is the GCF of 45 and 75? _____ .

3.   Circle each number below whose only factors are one and itself.

      2   3   4   5   6   7   8   9   10   11   12   13

During Class:

☐   **Write your class notes.** Neatly write down **all** examples shown as well as key terms or phrases with definitions. If not applicable or if you were absent, watch the Lecture Series (DVD) for this section and do the same (write down the examples shown as well as key terms or phrases). Insert more paper as needed.

Class Notes/Examples	**Your Notes**

Answers: **1)** smallest   **2)** for 45: $1, 3, 5, 9, 15, 45$; for 75: $1, 3, 5, 15, 25, 75$; GCF: 15
3) 2, 3, 5, 7, 11, and 13

Section 13.1 The Greatest Common Factor

Class Notes (continued)	**Your Notes**

(Insert additional paper as needed.)

Section 13.1 The Greatest Common Factor

Practice:

☐   Complete the Vocabulary and Readiness Check on page 982.

☐   Next, complete any incomplete exercises below. Check and correct your work using the answers and references at the end of this section.

Review this example:

1. Find the GCF for each list.

   a. 28, 40

   b. $6x^2, 10x^3,$ and $-8x$

a. Write each as a product of primes.

$28 = \boxed{2 \cdot 2} \cdot 7 = 2^2 \cdot 7$
$40 = \boxed{2 \cdot 2} \cdot 2 \cdot 5 = 2^3 \cdot 5$

There are two common factors, GCF = $2 \cdot 2 = \boxed{4}$

b. Write each as a product of primes.

$6x^2 = \boxed{2} \cdot 3 \cdot x^2$
$10x^3 = \boxed{2} \cdot 5 \cdot x^3$
$-8x = -1 \cdot \boxed{2} \cdot 2 \cdot 2 \cdot x$

GCF = $2 \cdot x^1 = \boxed{2x}$

Remember, the GCF contains the smallest exponent on the common variable.

Your turn:

2. Find the GCF for each list.

   a. 36, 90

   b. $12y^4, 20y^3$

Complete this example:

3. Factor out the GCF from each polynomial.

   a. $5ab + 10a$

   b. $-9a^5 + 18a^2 - 3a$

a. The GCF of $5ab$ and $10a$ is $5a$.

$5ab + 10a = 5a \cdot b + 5a \cdot 2$ Factor using the GCF.

$= 5a(b+2)$ Apply the distributive property.

b. The GCF of $-9a^5$, $18a^2$, and $-3a$ is $3a$.

$-9a^5 + 18a^2 - 3a = 3a(-3a^4) + 3a(6a) + 3a(-1)$

$= 3a(\underline{})$

Your turn:

4. Factor out the GCF from each polynomial.

   a. $30x - 15$

   b. $14x^3y + 7x^2y - 7xy$

Section 13.1 The Greatest Common Factor

Review this example:

5. Factor: $5(x+3) + y(x+3)$

The binomial $(x+3)$ is present in both terms and is the greatest common factor. We use the distributive property to factor out $(x+3)$.

$$5(x+3) + y(x+3) = (x+3)(5+y)$$

Your turn:

6. Factor: $y(x^2+2) + 3(x^2+2)$

Complete this example:

7. Factor by grouping: $3x^2 + 4xy - 3x - 4y$

$3x^2 + 4xy - 3x - 4y$

$= (3x^2 + 4xy) + (-3x - 4y)$ Group the terms.

$= x(3x + 4y) - 1(3x + 4y)$ Factor each group.

$= (3x + 4y)(x - 1)$

Factor out the common factor.

Check: Multiply $(3x + 4y)(x - 1)$.

Your turn:

8. Factor by grouping.

$$5xy - 15x - 6y + 18$$

	Answer	Text Ref	Video Ref		Answer	Text Ref	Video Ref
1	a. 4 b. $2x$	Ex 1a, 3a, p. 976; 978		5	$(x+3)(5+y)$	Ex 9, p. 979	
2	a. 18 b. $4y^3$		Sec 13.1, 2, 4/9	6	$(x^2+2)(y+3)$		Sec 13.1, 7/9
3	a. $5a(b+2)$ b. $3a(-3a^4 + 6a - 1)$	Ex 4a, 5, p. 979		7	$(3x+4y)(x-1)$	Ex 12, p. 980	
4	a. $15(2x-1)$ b. $7xy(2x^2 + x - 1)$		Sec 13.1, 5-6/9	8	$(y-3)(5x-6)$		Sec 13.1, 8/9

☐   **Next, insert your homework.** Make sure you attempt all exercises asked of you and show all work, as in the exercises above. Check your answers if possible. Clearly mark any exercises you were unable to correctly complete so that you may ask questions later. DO NOT ERASE YOUR INCORRECT WORK. THIS IS HOW WE UNDERSTAND AND EXPLAIN TO YOU YOUR ERRORS.

Section 13.2 Factoring Trinomials of the Form $x^2 + bx + c$

Before Class:

☐   Read the objectives on page 986.

☐   Read the **Helpful Hint** boxes on pages 986 and 989.

☐   Complete the exercises:

1. $(x+2)(x+5) = (x+5)(x+2)$ since multiplication is _____.

2.   Find two numbers whose product is 10 and whose sum is 7.

3.   Find two numbers whose product is 10 and whose sum is -7.

4.   Find two numbers whose product is -10 and whose sum is 3.

5.   Find two numbers whose product is -10 and whose sum is -3.

During Class:

☐   **Write your class notes.** Neatly write down **all** examples shown as well as key terms or phrases with definitions. If not applicable or if you were absent, watch the Lecture Series (DVD) for this section and do the same (write down the examples shown as well as key terms or phrases). Insert more paper as needed.

Class Notes/Examples	**Your Notes**

Answers: **1)** commutative   **2)** 5 and 2   **3)** −5 and −2   **4)** 5 and −2   **5)** −5 and 2

Section 13.2 Factoring Trinomials of the Form $x^2 + bx + c$

Class Notes (continued)	**Your Notes**

(Insert additional paper as needed.)

Section 13.2 Factoring Trinomials of the Form $x^2 + bx + c$

Practice:

☐   Complete the Vocabulary and Readiness Check on page 990.

☐   Next, complete any incomplete exercises below. Check and correct your work using the
answers and references at the end of this section.

Review this example:
1. Factor each trinomial completely.

   a. $x^2 + 7x + 12$

   b. $r^2 - r - 42$

a. $x^2 + 7x + 12 = (x + \square)(x + \square)$

Find two numbers whose product is 12 and whose
sum is 7. The numbers are 3 and 4.

$x^2 + 7x + 12 = \boxed{(x+3)(x+4)}$

b. $r^2 - r - 42 = (r + \square)(r + \square)$

Find two numbers whose product is -42 and
whose sum is -1. The numbers are 6 and -7.

$r^2 - r - 42 = \boxed{(r+6)(r-7)}$

Your turn:
2. Factor each trinomial completely.

   a. $x^2 + 7x + 6$

   b. $x^2 - 3x - 18$

Review this example:
3. Factor: $x^2 + 5xy + 6y^2$

$x^2 + 5xy + 6y^2 = (x + \square y)(x + \square y)$

Find two numbers whose product is 6 and whose
sum is 5. The numbers are 2 and 3.

$x^2 + 5xy + 6y^2 = \boxed{(x+2y)(x+3y)}$

Your turn:
4. Factor: $x^2 - 3xy - 4y^2$

341

Section 13.2 Factoring Trinomials of the Form $x^2 + bx + c$

Complete this example:

5. Factor completely: $3m^2 - 24m - 60$

Factor out the greatest common factor from each term. The greatest common factor is 3.

$$3m^2 - 24m - 60 = 3(m^2 - 8m - 20)$$
$$= 3(m + \square)(m + \square)$$

The resulting trinomial, $m^2 - 8m - 20$, can be factored by finding two numbers whose product is -20 and whose sum is -8. The numbers are -10 and 2.

Write the completely factored form below:

$3m^2 - 24m - 60 =$

Your turn:

6. Factor completely: $3x^2 + 9x - 30$

	Answer	Text Ref	Video Ref		Answer	Text Ref	Video Ref
1	a. $(x+3)(x+4)$ b. $(r+6)(r-7)$	Ex 1, 4, p. 987 - 988		**4**	$(x-4y)(x+y)$		Sec 13.2, 4/6
2	a. $(x+6)(x+1)$ b. $(x-6)(x+3)$		Sec 13.2, 1,3/6	**5**	$3(m-10)(m+2)$	Ex 9, p. 989	
3	$(x+2y)(x+3y)$	Ex 6, p. 988		**6**	$3(x+5)(x-2)$		Sec 13.2, 5/6

☐ **Next, insert your homework.** Make sure you attempt all exercises asked of you and show all work, as in the exercises above. Check your answers if possible. Clearly mark any exercises you were unable to correctly complete so that you may ask questions later. DO NOT ERASE YOUR INCORRECT WORK. THIS IS HOW WE UNDERSTAND AND EXPLAIN TO YOU YOUR ERRORS.

Section 13.3 Factoring Trinomials of the Form $ax^2 + bx + c$

Before Class:

☐   Read the objectives on page 993.

☐   Read the **Helpful Hint** boxes on pages 994 and 996.

☐   Complete these exercises:

Fill in each blank with "true" or "false."

1. To factor $x^2 + 11x + 28$ we look for two numbers whose product is 28 and whose sum is 11. _____

2. We can write the factorization of $(x + 2)(x - 3)$ also as $(x - 3)(x + 2)$. _____

3. The factorization of $(2x + 4)(x - 3)$ is completely factored. _____

4. A polynomial that factors as $(x + 4y)(x - 4y)$ is $x^2 + 16y^2$. _____

During Class:

☐   **Write your class notes.** Neatly write down **all** examples shown as well as key terms or phrases with definitions. If not applicable or if you were absent, watch the Lecture Series (DVD) for this section and do the same (write down the examples shown as well as key terms or phrases). Insert more paper as needed.

Class Notes/Examples	**Your Notes**

Answers: **1)** true   **2)** true   **3)** false   **4)** false

Section 13.3 Factoring Trinomials of the Form $ax^2 + bx + c$

Class Notes (continued)	**Your Notes**

(Insert additional paper as needed.)

Section 13.3 Factoring Trinomials of the Form $ax^2 + bx + c$

Practice:

☐   Complete the Vocabulary and Readiness Check on page 997.

☐   Next, complete any incomplete exercises below. Check and correct your work using the answers and references at the end of this section.

Review this example:

1. Factor: $3x^2 + 11x + 6$

Find factors of $3x^2$: $3x \cdot x$.
Find factors of 6: $1 \cdot 6$, $2 \cdot 3$.

$(3x + 1)(x + 6)$ Try $1 \cdot 6$ first. Middle term $11x$?

$1x$

$18x$

$1x + 18x = 19x$, incorrect middle term

$(3x + 2)(x + 3)$ Try $2 \cdot 3$ first. Middle term $11x$?

$2x$

$9x$

$2x + 9x = 11x$, correct middle term.

The factored form is $(3x + 2)(x + 3)$.

Your turn:

2. Factor: $10x^2 + 31x + 3$

Review this example:

3. Factor: $2x^2 + 13x - 7$

Find factors of $2x^2$: $2x \cdot x$.
Find factors of -7: $1 \cdot -7$, $-1 \cdot 7$.

Try possible combinations of these factors:

$(2x + 1)(x - 7) = 2x^2 - 13x - 7$ Incorrect middle term
$(2x - 1)(x + 7) = 2x^2 + 13x - 7$ Correct middle term

The factored form is $(2x - 1)(x + 7)$.

Your turn:

4. Factor: $4x^2 - 8x - 21$

345

Section 13.3 Factoring Trinomials of the Form $ax^2 + bx + c$

Review this example:

5. Factor: $24x^4 + 40x^3 + 6x^2$

Notice all three terms have a common factor: $2x^2$.

$= 2x^2(12x^2 + 20x + 3)$ Factor out $2x^2$ first.

Find factors of $12x^2$: $4x \cdot 3x$, $12x \cdot x$, $2x \cdot 6x$.

Find factors of 3: $1 \cdot 3$.

Try possible combinations of these factors:

$2x^2(4x+3)(3x+1) = 2x^2(12x^2 + 13x + 3)$ ☒

$2x^2(12x+1)(x+3) = 2x^2(12x^2 + 37x + 3)$ ☒

$2x^2(2x+3)(6x+1) = 2x^2(12x^2 + 20x + 3)$ ✓

The factored form is $2x^2(2x+3)(6x+1)$.

Your turn:

6. Factor: $4x^3 - 9x^2 - 9x$

Complete this example:

7. Factor: $-6x^2 - 13x + 5$

We begin by factoring out a common factor of -1.

$= -1(6x^2 + 13x - 5)$

Find factors of $6x^2$: $2x \cdot 3x$, $1x \cdot 6x$.

Find factors of -5: $-1 \cdot 5$, $1 \cdot -5$.

Try possible combinations of these factors:

$-1(2x-1)(3x+5) = -6x^2 - 7x + 5$ ☒

$-1(3x-1)(2x+5) = $ _____   ✓

The factored form is ⬭

Your turn:

8. Factor: $-14x^2 + 39x - 10$

	Answer	Text Ref	Video Ref		Answer	Text Ref	Video Ref
1	$(3x+2)(x+3)$	Ex 1, p. 994		**5**	$2x^2(2x+3)(6x+1)$	Ex 6, p. 996	
2	$(10x+1)(x+3)$		Sec 13.3, 1/6	**6**	$x(4x+3)(x-3)$		Sec 13.3, 4/6
3	$(2x-1)(x+7)$	Ex 3, p. 995		**7**	$-1(3x-1)(2x+5)$	Ex 7, p. 996	
4	$(2x-7)(2x+3)$		Sec 13.3, 2/6	**8**	$-1(2x-5)(7x-2)$		Sec 13.3, 5/6

☐   **Next, insert your homework.** Make sure you attempt all exercises asked of you and show all work, as in the exercises above. Check your answers if possible. Clearly mark any exercises you were unable to correctly complete so that you may ask questions later. DO NOT ERASE YOUR INCORRECT WORK. THIS IS HOW WE UNDERSTAND AND EXPLAIN TO YOU YOUR ERRORS.

Section 13.4 Factoring Trinomials of the Form $ax^2 + bx + c$ by Grouping

Before Class:

☐ Read the objective on page 1000.

☐ Complete the exercises:

1. The trinomial $2x^2 - 3x + 4$ is in the form $ax^2 + bx + c$. State the values for a, b, and c.

2. Factor out the GCF: $3x^2 - 6x$

3. Factor out the GCF: $3x^3 + 6x^2 - 24x$

4. Factor $8x^2 - 24x - 5x + 15$ by grouping.

During Class:

☐ **Write your class notes.** Neatly write down **all** examples shown as well as key terms or phrases with definitions. If not applicable or if you were absent, watch the Lecture Series (DVD) for this section and do the same (write down the examples shown as well as key terms or phrases). Insert more paper as needed.

Class Notes/Examples	**Your Notes**

Answers: **1)** $a = 2, b = -3$ and $c = 4$ **2)** $3x(x - 2)$ **3)** $3x(x^2 + 2x - 8)$ **4)** $(x - 3)(8x - 5)$

Section 13.4 Factoring Trinomials of the Form $ax^2 + bx + c$ by Grouping

Class Notes (continued)

Your Notes

(Insert additional paper as needed.)

Section 13.4 Factoring Trinomials of the Form $ax^2 + bx + c$ by Grouping

Practice:

☐   Complete the Vocabulary and Readiness Check on page 1002.

☐   Next, complete any incomplete exercises below. Check and correct your work using the answers and references at the end of this section.

Review this example:	**Your turn:**
1. Factor $8x^2 - 14x + 5$ by grouping.	**2.** Factor $10x^2 - 9x + 2$ by grouping.

Step 1. Factor out the greatest common factor.

   The GCF = 1.

Step 2. Find two numbers whose product is $a \cdot c$ and whose sum is b.

   $8x^2 - 14x + 5$

   a = 8, b = -14, and c = 5

   Find two numbers whose product is 40 and whose sum is -14. The numbers are -4 and -10.

Step 3. Write the middle term $-14x$, using the numbers -4 and -10.

   $8x^2 - 14x + 5 = 8x^2 - 4x - 10x + 5$

Step 4. Factor by grouping.

   $8x^2 - 4x - 10x + 5 = 4x(2x-1) - 5(2x-1)$

   $= (2x-1)(4x-5)$

349

Section 13.4 Factoring Trinomials of the Form $ax^2 + bx + c$ by Grouping

Complete this example:

3. Factor $6x^2 - 2x - 20$ by grouping.

Step 1. Factor out the GCF. The GCF = 2.

$$2(3x^2 - x - 10)$$

Step 2. In the resulting trinomial $3x^2 - x - 10$, find two numbers whose product is -30 and whose sum is -1. The numbers are -6 and 5.

Step 3. Write the middle term $-x$, using the numbers -6 and 5.

$$3x^2 - x - 10 = 3x^2 - 6x + 5x - 10$$

Step 4. Factor by grouping. Remember to include the GCF in your final answer.

$$3x^2 - 6x + 5x - 10$$

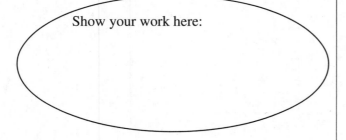

Show your work here:

Your turn:

4. Factor: $12x^3 - 27x^2 - 27x$

	Answer	Text Ref	Video Ref		Answer	Text Ref	Video Ref
1	$(2x-1)(4x-5)$	Ex 1, p. 1001		3	$2(x-2)(3x+5)$	Ex 2, p. 1001	
2	$(5x-2)(2x-1)$		Sec 11.4, 3/4	4	$3x(4x+3)(x-3)$		Sec 11.4, 4/4

☐ **Next, insert your homework.** Make sure you attempt all exercises asked of you and show all work, as in the exercises above. Check your answers if possible. Clearly mark any exercises you were unable to correctly complete so that you may ask questions later. DO NOT ERASE YOUR INCORRECT WORK. THIS IS HOW WE UNDERSTAND AND EXPLAIN TO YOU YOUR ERRORS.

Section 13.5 Factoring Perfect Square Trinomials and the Difference of Two Squares

Before Class:

☐   Read the objectives on page 1004.

☐   Read the **Helpful Hint** boxes on pages 1004, 1006 and 1007.

☐   Complete these exercises:

Find all positive values of b so that each trinomial is factorable.

1. $3x^2 + bx - 5$ 2. $2x^2 + bx - 7$

Find all positive values of c so that each trinomial is factorable.

3. $5x^2 + 7x + c$ 4. $3x^2 - 8x + c$

During Class:

☐   **Write your class notes.** Neatly write down **all** examples shown as well as key terms or phrases with definitions. If not applicable or if you were absent, watch the Lecture Series (DVD) for this section and do the same (write down the examples shown as well as key terms or phrases). Insert more paper as needed.

Class Notes/Examples	**Your Notes**

Answers: **1)**   2; 14   **2)**   5; 13   **3)**   2   **4)**   4; 5

Section 13.5 Factoring Perfect Square Trinomials and the Difference of Two Squares

Class Notes (continued)	**Your Notes**

(Insert additional paper as needed.)

Section 13.5 Factoring Perfect Square Trinomials and the Difference of Two Squares

Practice:

☐   Complete the Vocabulary and Readiness Check on page 1009.

☐   Next, complete any incomplete exercises below. Check and correct your work using the answers and references at the end of this section.

Review this example: **1.** Determine whether the given trinomial is a perfect square trinomial: $x^2 + 8x + 16$ Step 1. The first term, $x^2 = (x)^2$, and the last term, $16 = (4)^2$, are squares. Step 2. $2 \cdot x \cdot 4 = 8x$, the middle term. (YES)	**Your turn:** **2.** Determine whether each trinomial is a perfect square trinomial. a. $x^2 + 16x + 64$ b. $y^2 + 5y + 25$
Review this example: **3.** Factor the trinomial $x^2 + 12x + 36$ completely. Notice that every term is positive. Check to see if the trinomial is of the form $a^2 + 2ab + b^2$. $$x^2 + 12x + 36 = x^2 + 2 \cdot x \cdot 6 + 6^2$$ $$\uparrow \quad \uparrow \uparrow \uparrow \uparrow$$ $$a^2 + 2 \cdot a \cdot b + b^2$$ $$= (x+6)^2$$ $$\uparrow \uparrow \uparrow$$ $$(a+b)^2$$	**Your turn:** **4.** Factor the trinomial completely: $x^2 + 22x + 121$
Review this example: **5.** Factor the trinomial $4m^4 - 4m^2 + 1$. Notice that the middle term is negative. Check to see if the trinomial is of the form $a^2 - 2ab + b^2$. $$4m^4 - 4m^2 + 1 = (2m^2)^2 - 2 \cdot 2m^2 \cdot 1 + 1^2$$ $$\uparrow \quad \uparrow \uparrow \quad \uparrow \uparrow$$ $$a^2 \quad -2 \cdot a \cdot b \ + b^2$$ $$= (2m^2 - 1)^2$$ $$\uparrow \quad \uparrow \uparrow$$ $$(a \ - b)^2$$	**Your turn:** **6.** Factor the trinomial completely: $9x^2 - 24xy + 16y^2$

Section 13.5 Factoring Perfect Square Trinomials and the Difference of Two Squares

Review this example:

7. Factor each binomial completely:

   a. $y^2 - 25$ b. $x^2 + 4$

Check the binomial to see if:
1. both terms are squares.
2. the signs of the terms are different.

a. $y^2 - 25 = (y)^2 - 5^2 = (y+5)(y-5)$

b. $x^2 + 4 = (x)^2 + 2^2$ Notice the signs are NOT
                 different.   PRIME

Common mistakes are:
$(x+2)(x+2) = x^2 + 4x + 4$ Results in trinomial.
$(x-2)(x-2) = x^2 - 4x + 4$ Results in trinomial.

Your turn:

8. Factor each binomial completely.

   a. $x^2 - 4$

   b. $16r^2 + 1$

Complete this example:

9. Factor each binomial completely:

   a. $4x^3 - 49x$ b. $-49x^2 + 16$

a. $= x(4x^2 - 49)$ Factor out common factor, x.

   $= x[(2x)^2 - 7^2]$ Factor the difference

   $= x(2x+7)(2x-7)$ of two squares.

b. $= -1(49x^2 - 16)$ Factor out -1.

   $= -1(\_\_\_ + \_\_\_)(\_\_\_ - \_\_\_)$

Your turn:

10. Factor each binomial completely.

   a. $xy^3 - 9xyz^2$

   b. $121m^2 - 100n^2$

	Answer	Text Ref		Answer	Video Ref
1	yes	Ex 1, p. 1004	2	a. yes b. no	Sec 11.5, 1,2/9
3	$(x+6)^2$	Ex 4, p. 1005	4	$(x+11)^2$	Sec 11.5, 3/9
5	$(2m^2 - 1)^2$	Ex 6, p. 1005	6	$(3x - 4y)^2$	Sec 11.5, 4/9
7	a. $(y+5)(y-5)$ b. Prime	Ex 10&12, p. 1006	8	a. $(x+2)(x-2)$ b. Prime	Sec 11.5, 5,7/9
9	a. $x(2x+7)(2x-7)$ b. $-1(7x+4)(7x-4)$	Ex 16, 18 p. 1007	10	a. $xy(y-3z)(y+3z)$ b. $(11m+10n)(11m-10n)$	Sec 11.5, 8,6/9

☐   **Next, insert your homework.** Make sure you attempt all exercises asked of you and show all work, as in the exercises above. Check your answers if possible. Clearly mark any exercises you were unable to correctly complete so that you may ask questions later. DO NOT ERASE YOUR INCORRECT WORK. THIS IS HOW WE UNDERSTAND AND EXPLAIN TO YOU YOUR ERRORS.

Section 13.6 Solving Quadratic Equations by Factoring

Before Class:

☐   Read the objectives on page 1015.

☐   Read the **Helpful Hint** boxes on pages 1016 and 1018.

☐   Read the definition of the standard form of a quadratic equation and the definition of the zero-factor property on page 1015.

☐   Complete the exercises:

1.   The zero-factor property says that if a product is 0, then a factor = _____ .

2.   If $(x-1)(x+1) = 0$, then by the zero-factor property $x - 1 = $ ___ or $x + 1 = $ ___ .

3.   Write $2x^2 - 7x = 4$ in standard form.

4.   Factor: $x^2 - 9$

5.   Factor: $x^2 - 5x + 6$

During Class:

☐   **Write your class notes.** Neatly write down **all** examples shown as well as key terms or phrases with definitions. If not applicable or if you were absent, watch the Lecture Series (DVD) for this section and do the same (write down the examples shown as well as key terms or phrases). Insert more paper as needed.

Class Notes/Examples	**Your Notes**

Answers: **1)** zero   **2)** both = 0   **3)** $2x^2 - 7x - 4 = 0$ **4)** $(x+3)(x-3)$ **5)** $(x-2)(x-3)$

Section 13.6 Solving Quadratic Equations by Factoring

Class Notes (continued)	Your Notes

(Insert additional paper as needed.)

Section 13.6 Solving Quadratic Equations by Factoring

Practice:

☐   Complete the Vocabulary and Readiness Check on page 1020.

☐   Next, complete any incomplete exercises below. Check and correct your work using the answers and references at the end of this section.

Review this example:

1. Solve: $(x-5)(2x+7)=0$

The product is zero. Using the zero-factor property, set each factor equal to zero and solve each linear equation.

$(x-5)(2x+7)=0$

$x-5=0$ $2x+7=0$

       or

$\boxed{x=5}$ $\boxed{x=-\dfrac{7}{2}}$

Check $x=5$:   Check $x=-\dfrac{7}{2}$:

$(5-5)(2x+7)\overset{?}{=}0$

$0\cdot(2x+7)\overset{?}{=}0$ $(x-5)\left(\cancel{2}\left(-\dfrac{7}{\cancel{2}}\right)+7\right)\overset{?}{=}0$

$0=0$ $(x-5)\cdot0\overset{?}{=}0$

Your turn:

2. Solve: $(2x+3)(4x-5)=0$

Review this example:

3. Solve: $x^2-9x-22=0$

$x^2-9x-22=0$ The product equals 0.

$(x-11)(x+2)=0$ Factor.

$x-11=0$ $x+2=0$ Set each factor equal to

       or   zero and solve.

$\boxed{x=11}$ $\boxed{x=-2}$

Check $x=11$:

$(11)^2-9(11)-22\Rightarrow121-99-22\Rightarrow22-22=0$

Check $x=-2$:

$(-2)^2-9(-2)-22\Rightarrow4+18-22\Rightarrow22-22=0$

Your turn:

4. Solve: $x^2+2x-8=0$

Section 13.6 Solving Quadratic Equations by Factoring

Review this example:	**Your turn:**

Review this example:

5. Solve: $x(5x-2)=0$

$x(5x-2)=0$ The product = zero.

 $x=0$ or $5x-2=0$ Set each factor equal to zero and solve.

$x=\dfrac{2}{5}$

Your turn:

6. Solve: $4y^2-1=0$

Complete this example:

7. Solve: $x(2x-7)=4$

$x(2x-7)=4$ The product $\neq$ zero.

$2x^2-7x=4$ Distribute.

$2x^2-7x-4=0$ Write in standard form.

$2x^2-8x+x-4=0$ Factor by grouping.

$2x(\quad\quad)+1(\quad\quad)=0$

$(2x+1)(\quad\quad)=0$

$2x+1=0$ or $\bigcirc\;=0$ Set each factor equal to zero.

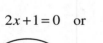 $x=$ or $x=$ Solve.

Your turn:

8. Solve: $x(3x-1)=14$

	Answer	Text Ref	Video Ref		Answer	Text Ref	Video Ref
1	$x=5,\ x=-\dfrac{7}{2}$	Ex 2, p. 1016		5	$x=0,\ x=\dfrac{2}{5}$	Ex 3, p. 1017	
2	$x=-\dfrac{3}{2},\ x=\dfrac{5}{4}$		Sec 11.6, 2/4	6	$y=\dfrac{1}{2},\ y=-\dfrac{1}{2}$		Not a video
3	$x=11,\ x=-2$	Ex 4, p. 1017		7	$(x-4),\ x=-\dfrac{1}{2},\ x=4$	Ex 6, p. 1018	
4	$x=-4,\ x=2$		Sec 11.6, 1/4	8	$x=\dfrac{7}{3},\ x=-2$		Sec 11.6, 3/4

☐   **Next, insert your homework.** Make sure you attempt all exercises asked of you and show all work, as in the exercises above. Check your answers if possible. Clearly mark any exercises you were unable to correctly complete so that you may ask questions later. DO NOT ERASE YOUR INCORRECT WORK. THIS IS HOW WE UNDERSTAND AND EXPLAIN TO YOU YOUR ERRORS.

Section 13.7 Quadratic Equations and Problem Solving

Before Class:

☐   Read the objective on page 1023.

☐   Read the **Helpful Hint** box and the Pythagorean Theorem on page 1026.

☐   Complete the exercises:

1.   Factor completely: $-2t^2 + 8$

2.   Write $x^2 + 3x = 70$ in standard form.

3.   Is $x = 7$ a solution to the quadratic equation $x^2 + 3x = 70$?

4.   Let x be an even integer. Write the next two consecutive even integers in terms of x.

During Class:

☐   **Write your class notes.** Neatly write down **all** examples shown as well as key terms or phrases with definitions. If not applicable or if you were absent, watch the Lecture Series (DVD) for this section and do the same (write down the examples shown as well as key terms or phrases). Insert more paper as needed.

Class Notes/Examples	Your Notes

Answers: **1)** $-2(t+2)(t-2)$ **2)** $x^2 + 3x - 70 = 0$ **3)** yes; $49 + 21 = 70$ **4)** $x+2, x+4$

Section 13.7 Quadratic Equations and Problem Solving

Class Notes (continued)	Your Notes

(Insert additional paper as needed.)

Section 13.7 Quadratic Equations and Problem Solving

Practice:

☐   Next, complete any incomplete exercises below. Check and correct your work using the answers and references at the end of this section.

Review this example:

1. Neglecting air resistance, the height h in feet of a cliff diver above the ocean after t seconds is given by the quadratic equation $h = -16t^2 + 144$. Find out how long it takes the diver to reach the water.

When the diver reaches the water the height is zero. To find out how long it takes for the diver to reach the water, solve the equation for t when $h = 0$.

$$-16t^2 + 144 = 0 \qquad \text{Let } h = 0.$$

$$-16(t^2 - 9) = 0 \qquad \text{Factor out } -16.$$

$$-16(t + 3)(t - 3) = 0 \qquad \text{Factor completely.}$$

$t + 3 = 0 \quad$ or $\quad t - 3 = 0$ Set each factor containing t equal to zero and solve.
$\quad t = -3 \qquad\qquad t = 3$

$t = -3 \ $ or $\ \boxed{t = 3}$
      ↑

Time cannot be negative. Therefore, it takes the diver 3 seconds to reach the water.

Your turn:

2. An object is thrown upward from the top of an 80-foot building with an initial velocity of 64 feet per second. The height h of the object after t seconds is given by the quadratic equation $h = -16t^2 + 64t + 80$. When will the object hit the ground?

Review this example:

3. The square of a number plus three times the number is 70. Find the number.

Let $x =$ the number.
Translate and solve: $x^2 + 3x = 70$

$$x^2 + 3x = 70 \qquad \text{Write in standard form.}$$

$$x^2 + 3x - 70 = 0 \qquad \text{Factor and solve.}$$

$$(x + 10)(x - 7) = 0$$

$$x + 10 = 0 \quad \text{or} \quad x - 7 = 0$$

$$\boxed{x = -10 \ \text{or} \qquad x = 7}$$

Your turn:

4. The sum of a number and its square is 182. Find the number.

Section 13.7 Quadratic Equations and Problem Solving

Complete this example:

5. Find the lengths of the sides of a right triangle if the lengths can be expressed as three consecutive even integers.

Let x, $x+2$, and $x+4$ be three consecutive even integers. The longest side, $x+4$, is the hypotenuse. The other two sides are the legs.

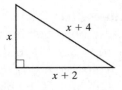

By the Pythagorean Theorem,

$$(x)^2 + (x+2)^2 = (x+4)^2$$

$x^2 + x^2 + 4x + 4 = x^2 + 8x + 16$ Multiply.

$2x^2 + 4x + 4 = x^2 + 8x + 16$ Simplify.

Next, write the equation in standard form.

Factor and solve the quadratic equation.

Show your work here:

Your turn:

6. One leg of a right triangle is 4 millimeters longer than the smaller leg and the hypotenuse is 8 millimeters longer than the smaller leg. Find the lengths of the sides of the triangle.

	Answer	Text Ref	Video Ref		Answer	Text Ref	Video Ref
1	3 seconds	Ex 1, p. 1023		4	$x=-14$, $x=13$		Sec 11.7, 4/5
2	5 seconds		Sec 11.7, 2/5	5	$x^2-4x-12=0$ 6 units, 8 units, 10 units	Ex 5, p. 1026	
3	$x=-10$, $x=7$	Ex 2, p. 1024		6	Let x = the smaller leg. 12 mm, 16 mm, 20 mm		Sec 11.7, 3/5

☐   **Next, insert your homework.** Make sure you attempt all exercises asked of you and show all work, as in the exercises above. Check your answers if possible. Clearly mark any exercises you were unable to correctly complete so that you may ask questions later. DO NOT ERASE YOUR INCORRECT WORK. THIS IS HOW WE UNDERSTAND AND EXPLAIN TO YOU YOUR ERRORS.

Preparing for the Chapter 13 Test

Start preparing for your Chapter 13 Test as soon as possible. Pay careful attention to any instructor discussion about this test, especially discussion on what sections you will be responsible for, etc.

☐ Work the Chapter 13 Vocabulary Check on page 1033.

☐ Read both columns (Definitions and Concepts, and Examples) of the Chapter 13 Highlights starting on page 1033.

☐ Read your Class Notes/Examples for each section covered on your Chapter 13 Test. Look for any unresolved questions you may have.

☐ Complete as many of the Chapter 13 Review exercises as possible (starting on page 1036). Remember, the odd answers are in the back of your text.

☐ **Most important:** Place yourself in "test" conditions (see below) and work the Chapter 13 Test (pages 1040 - 1041) as a practice test the day before your actual test. To honestly assess how you are doing, try the following:
   - Work on a few blank sheets of paper.
   - Give yourself the same amount of time you will be given for your actual test.
   - Complete this Chapter 13 Practice Test without using your notes or your text.
   - If you have any time left after completing this practice test, check your work and try to find any errors on your own.
   - Once done, use the back of your book to check ALL answers.
   - Try to correct any errors on your own.
   - Use the Chapter Test Prep Video (CTPV) to correct any errors you were unable to correct on your own. You can find these videos in the Interactive DVD Lecture Series, in MyMathLab, and on YouTube. Search MartinGayDevMath and click "Channels."

I wish you the best of luck….Elayn Martin-Gay

Section 14.1 Simplifying Rational Expressions

Before Class:

☐ Read the objectives on page 1046.

☐ Read the **Helpful Hint** boxes on pages 1047, 1049, and 1051.

☐ Complete the exercises:

1. Factor: $6x^2 + 6x$

2. Factor: $9x^2 - 16$

3. Factor: $x^2 - 3x + 2$

4. Divide out common factors: $\dfrac{9(x+1)}{18(x+1)}$

During Class:

☐ **Write your class notes.** Neatly write down **all** examples shown as well as key terms or phrases with definitions. If not applicable or if you were absent, watch the Lecture Series (DVD) for this section and do the same (write down the examples shown as well as key terms or phrases). Insert more paper as needed.

Class Notes/Examples	Your Notes

Answers: **1)** $6x(x+1)$ **2)** $(3x-4)(3x+4)$ **3)** $(x-2)(x-1)$ **4)** $\dfrac{1}{2}$

Section 14.1 Simplifying Rational Expressions

Class Notes (continued)	Your Notes

(Insert additional paper as needed.)

Section 14.1 Simplifying Rational Expressions

Practice:

☐   Complete the Vocabulary and Readiness Check on page 1052.

☐   Next, complete any incomplete exercises below. Check and correct your work using the answers and references at the end of this section.

Review this example:

1. Find the value $\dfrac{x+4}{2x-3}$ when $x = 5$.

We replace each x in the expression with 5.

$$\frac{x+4}{2x-3} = \frac{5+4}{2(5)-3} = \frac{9}{10-3} = \boxed{\frac{9}{7}}$$

Your turn:

2. Find the value of $\dfrac{z}{z^2-5}$ when $z = -5$.

Complete this example:

3. Find any numbers for which each rational expression is undefined.

   a. $\dfrac{x}{x-3}$ b. $\dfrac{x^2+2}{x^2-3x+2}$

To find values for which a rational expression is undefined, we find values that make the denominator 0.

a. Set the denominator $x - 3 = 0$.

The expression is undefined, when $\boxed{x = 3}$

b. Set the denominator $x^2 - 3x + 2 = 0$.

$(x-2)(x-1) = 0$ Factor.

Set each factor = 0.

$(x-2) = 0$ or $(x-1) = 0$

    $x = \underline{}$ $x = \underline{}$

The expression is undefined, when

$\boxed{x = }$ or $\boxed{x = }$.

Your turn:

4. Find any numbers for which each rational expression is undefined.

   a. $\dfrac{x+3}{x+2}$

   b. $\dfrac{11x^2+1}{x^2-5x-14}$

Section 14.1 Simplifying Rational Expressions

Review this example:

5. Simplify:

a. $\dfrac{5x-5}{x^3-x^2}$ b. $\dfrac{x-y}{y-x}$

a. Factor each numerator and denominator.

$\dfrac{5x-5}{x^3-x^2} = \dfrac{5(x-1)}{x^2(x-1)} = \dfrac{5}{x^2}$ Divide out common factors.

b. Recognize that $y-x = -1(x-y)$.

$\dfrac{x-y}{y-x} = \dfrac{1\cdot(x-y)}{(-1)(x-y)} = \dfrac{1}{-1} = -1$ Divide out common factors.

Your turn:

6. Simplify:

a. $\dfrac{x-7}{7-x}$

b. $\dfrac{-5a-5b}{a+b}$

Review this example:

7. Simplify: $\dfrac{x+9}{x^2-81}$

Factor each numerator and denominator.
Divide out common factors.

$\dfrac{x+9}{x^2-81} = \dfrac{x+9}{(x+9)(x-9)} = \dfrac{1}{x-9}$

Your turn:

8. Simplify: $\dfrac{x^3+7x^2}{x^2+5x-14}$

	Answer	Text Ref	Video Ref		Answer	Text Ref	Video Ref
1	$\dfrac{9}{7}$	Ex 1a, p. 1046		5	a. $\dfrac{5}{x^2}$ b. -1	Ex 3, 7b, p. 1049,1050	
2	$-\dfrac{1}{4}$		Sec 14.1, 1/9	6	a. -1 b. -5		Sec 14.1, 6, 4/9
3	a. $x=3$ b. $x=2$ or $x=1$	Ex 2ab, p. 1047		7	$\dfrac{1}{x-9}$	Ex 6, p. 1050	
4	a. $x=-2$ b. $x=7$ or $x=-2$		Sec 14.1, 2–3/9	8	$\dfrac{x^2}{x-2}$		Sec 14.1, 7/9

☐ **Next, insert your homework.** Make sure you attempt all exercises asked of you and show all work, as in the exercises above. Check your answers if possible. Clearly mark any exercises you were unable to correctly complete so that you may ask questions later. DO NOT ERASE YOUR INCORRECT WORK. THIS IS HOW WE UNDERSTAND AND EXPLAIN TO YOU YOUR ERRORS.

Section 14.2 Multiplying and Dividing Rational Expressions

Before Class:

☐   Read the objectives on page 1056.

☐   Read the Steps used to Multiply Rational Expressions on page 1057.

☐   Read the **Helpful Hint** boxes on pages 1058 and 1061.

☐   Complete the exercises:

1.   Factor: $5x^2 - 5x$

2.   Factor: $4x^2 - 9$

3.   Factor: $2x^2 + x - 3$

4.   Divide out common factors: $\dfrac{x(x+1)\cdot 2\cdot 3}{3\cdot x\cdot 5(x+1)}$

During Class:

☐   **Write your class notes.** Neatly write down **all** examples shown as well as key terms or phrases with definitions. If not applicable or if you were absent, watch the Lecture Series (DVD) for this section and do the same (write down the examples shown as well as key terms or phrases). Insert more paper as needed.

Class Notes/Examples	**Your Notes**

Answers: **1)** $5x(x-1)$ **2)** $(2x-3)(2x+3)$ **3)** $(2x+3)(x-1)$ **4)** $\dfrac{2}{5}$

369

Section 14.2 Multiplying and Dividing Rational Expressions

Class Notes (continued)

Your Notes

(Insert additional paper as needed.)

Section 14.2 Multiplying and Dividing Rational Expressions

Practice:

☐   Complete the Vocabulary and Readiness Check on page 1062.

☐   Next, complete any incomplete exercises below. Check and correct your work using the answers and references at the end of this section.

Review this example:	**Your turn:**

1. Multiply and simplify: $\dfrac{-7x^2}{5y} \cdot \dfrac{3y^5}{14x^2}$

2. Multiply and simplify: $\dfrac{8x}{2} \cdot \dfrac{x^5}{4x^2}$

$\dfrac{-7x^2}{5y} \cdot \dfrac{3y^5}{14x^2} = \dfrac{-7x^2 \cdot 3y^5}{5y \cdot 14x^2}$ Multiply the numerators and denominators.

$= \dfrac{-1 \cdot 7 \cdot 3 \cdot x^2 \cdot y \cdot y^4}{5 \cdot 2 \cdot 7 \cdot x^2 \cdot y}$ Factor the numerator and denominator.

$= \dfrac{-1 \cdot \cancel{7} \cdot 3 \cdot \cancel{x^2} \cdot \cancel{y} \cdot y^4}{5 \cdot 2 \cdot \cancel{7} \cdot \cancel{x^2} \cdot \cancel{y}}$ Divide out common factors.

$= \dfrac{-3y^4}{10}$

Review this example:

3. Multiply and simplify:

$\dfrac{3x+3}{5x^2-5x} \cdot \dfrac{2x^2+x-3}{4x^2-9}$

Factor each numerator and denominator.

$\dfrac{3x+3}{5x^2-5x} \cdot \dfrac{2x^2+x-3}{4x^2-9} = \dfrac{3(x+1)}{5x(x-1)} \cdot \dfrac{(2x+3)(x-1)}{(2x-3)(2x+3)}$

Multiply the numerators and denominators.

$= \dfrac{3(x+1)(2x+3)(x-1)}{5x(x-1)(2x-3)(2x+3)}$

$= \dfrac{3(x+1)\cancel{(2x+3)}\cancel{(x-1)}}{5x\cancel{(x-1)}(2x-3)\cancel{(2x+3)}}$ Divide out common factors.

$= \dfrac{3(x+1)}{5x(2x-3)}$

Your turn:

4. Multiply and simplify:

$\dfrac{5x-20}{3x^2+x} \cdot \dfrac{3x^2+13x+4}{x^2-16}$

Section 14.2 Multiplying and Dividing Rational Expressions

Review this example:

5. Divide and simplify: $\dfrac{6x+2}{x^2-1} \div \dfrac{3x^2+x}{x-1}$

$\dfrac{6x+2}{x^2-1} \div \dfrac{3x^2+x}{x-1} = \dfrac{6x+2}{x^2-1} \cdot \dfrac{x-1}{3x^2+x}$ Multiply by the reciprocal.

$= \dfrac{2\cancel{(3x+1)}\,\cancel{(x-1)}}{(x+1)\,\cancel{(x-1)} \cdot x\cancel{(3x+1)}}$ Factor and multiply. Then divide out common factors.

$= \boxed{\dfrac{2}{x(x+1)}}$

Your turn:

6. Divide and simplify:

$\dfrac{5x-10}{12} \div \dfrac{4x-8}{8}$

Complete this example:

7. 18 square feet = _____ square yards.

From the diagram, you see that 1 square yard equals 9 square feet.

unit fraction

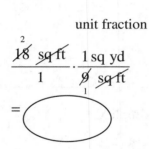

$\dfrac{\overset{2}{\cancel{18}}\ \cancel{sq\ ft}}{1} \cdot \dfrac{1\ sq\ yd}{\underset{1}{\cancel{9}}\ \cancel{sq\ ft}}$

$= \bigcirc$

1 yd = 3 ft

1 yd = 3 ft

Area: 1 sq yd or 9 sq ft

Your turn:

8. 3 cubic yards = _____ cubic feet

From the diagram, you see that 1 cubic yard equals 27 cubic feet.

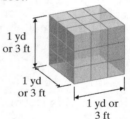

1 yd or 3 ft

1 yd or 3 ft

1 yd or 3 ft

	Answer	Text Ref	Video Ref		Answer	Text Ref	Video Ref
1	$-\dfrac{3y^4}{10}$	Ex 1b, p. 1056		**5**	$\dfrac{2}{x(x+1)}$	Ex 6, p. 1059	
2	x^4		Sec 14.2, 1/6	**6**	$\dfrac{5}{6}$		Sec 14.2, 4/6
3	$\dfrac{3(x+1)}{5x(2x-3)}$	Ex 3, p. 1057		**7**	2	Ex 9, p. 1060	
4	$\dfrac{5}{x}$		Sec 14.2, 2/6	**8**	81		Sec 14.2, 6/6

☐   **Next, insert your homework.** Make sure you attempt all exercises asked of you and show all work, as in the exercises above. Check your answers if possible. Clearly mark any exercises you were unable to correctly complete so that you may ask questions later. DO NOT ERASE YOUR INCORRECT WORK. THIS IS HOW WE UNDERSTAND AND EXPLAIN TO YOU YOUR ERRORS.

Section 14.3 Adding and Subtracting Rational Expressions

Before Class:

☐ Read the objectives on page 1065.

☐ Read the Adding and Subtracting Rational Expressions with Common Denominators box on page 1065 and the Find the Least Common Denominator box on page 1066.

☐ Read the **Helpful Hint** box on page 1066.

☐ Complete the exercises:

Some of the denominators will need to be factored first.

1. Find the area of the rectangle.

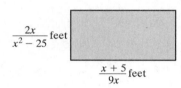

2. Find the area of the square.

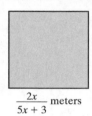

During Class:

☐ **Write your class notes.** Neatly write down **all** examples shown as well as key terms or phrases with definitions. If not applicable or if you were absent, watch the Lecture Series (DVD) for this section and do the same (write down the examples shown as well as key terms or phrases). Insert more paper as needed.

Class Notes/Examples	**Your Notes**

Answers: **1)** $\dfrac{2}{9(x-5)}$ sq ft   **2)** $\dfrac{4x^2}{(5x+3)^2}$ sq m

Section 14.3 Adding and Subtracting Rational Expressions

Class Notes (continued)

Your Notes

(Insert additional paper as needed.)

Section 14.3 Adding and Subtracting Rational Expressions

Practice:

☐   Complete the Vocabulary and Readiness Check on page 1070.

☐   Next, complete any incomplete exercises below. Check and correct your work using the answers and references at the end of this section.

Review this example:	**Your turn:**
1. Add: $\dfrac{5m}{2n} + \dfrac{m}{2n}$	**2.** Add: $\dfrac{9}{3+y} + \dfrac{y+1}{3+y}$

$\dfrac{5m}{2n} + \dfrac{m}{2n} = \dfrac{5m+m}{2n}$ Add the numerators.

$\qquad\qquad = \dfrac{6m}{2n}$ Simplify; combine like terms.

$\qquad\qquad \boxed{= \dfrac{3m}{n}}$ Simplify.

Review this example:	**Your turn:**
3. Subtract: $\dfrac{3x^2+2x}{x-1} - \dfrac{10x-5}{x-1}$	**4.** Subtract: $\dfrac{2x+3}{x^2-x-30} - \dfrac{x-2}{x^2-x-30}$

Subtract the numerators.

$\dfrac{3x^2+2x}{x-1} - \dfrac{10x-5}{x-1} = \dfrac{3x^2+2x-(10x-5)}{x-1}$

Use the distributive property.

$\qquad = \dfrac{3x^2+2x-10x+5}{x-1}$

Combine like terms.

$\qquad = \dfrac{3x^2-8x+5}{x-1}$

Factor.

$\qquad = \dfrac{(x-1)(3x-5)}{x-1}$

$\qquad = \dfrac{(x-1)(3x-5)}{x-1}$

$\qquad \boxed{= 3x-5}$

Section 14.3 Adding and Subtracting Rational Expressions

Review this example:

5. Find the LCD of $\dfrac{6m^2}{3m+15}, \dfrac{2}{(m+5)^2}$.

Factor each denominator.

$3m+15 = 3(m+5)$

$(m+5)^2 = (m+5)^2$ Denominator already factored.

Greatest number of times factor of 3 appears: 1
Greatest number of times factor of $m+5$ appears in
any one denominator: 2.

LCD $= 3(m+5)^2$

Your turn:

6. Find the LCD of
$\dfrac{1}{3x+3}, \dfrac{8}{2x^2+4x+2}$.

Review this example:

5. Write an equivalent rational expression using
the given denominator.

$$\frac{7x}{2x+5} = \frac{}{6x+15}$$

First, factor the denominator on the right.

$\dfrac{7x}{2x+5} = \dfrac{}{3(2x+5)}$ Multiply the left by $\dfrac{3}{3} = 1$.

$\dfrac{7x}{2x+5} = \dfrac{7x}{2x+5} \cdot \dfrac{3}{3} = \dfrac{7x \cdot 3}{(2x+5) \cdot 3} = \dfrac{21x}{3(2x+5)}$

Your turn:

8. Write an equivalent rational
expression using the given
denominator.

$$\frac{9a+2}{5a+10} = \frac{}{5b(a+2)}$$

	Answer	Text Ref	Video Ref		Answer	Text Ref	Video Ref
1	$\dfrac{3m}{n}$	Ex 1, p. 1065		5	$3(m+5)^2$	Ex 6, p. 1067	
2	$\dfrac{y+10}{3+y}$		Sec 14.3, 2/7	6	$6(x+1)^2$		Sec 14.3, 5/7
3	$3x-5$	Ex 3, p. 1066		7	$21x$	Ex 9b, p. 1068	
4	$\dfrac{1}{x-6}$		Sec 14.3, 3/7	8	$9ab+2b$		Sec 14.3, 7/7

☐   **Next, insert your homework.** Make sure you attempt all exercises asked of you and show
all work, as in the exercises above. Check your answers if possible. Clearly mark any
exercises you were unable to correctly complete so that you may ask questions later. DO
NOT ERASE YOUR INCORRECT WORK. THIS IS HOW WE UNDERSTAND AND
EXPLAIN TO YOU YOUR ERRORS.

Section 14.4 Adding and Subtracting Fractions with Different Denominators

Before Class:

☐   Read the objectives on page 1074.

☐   Read the steps to Add or Subtract Rational Expressions with Different Denominators on page 1074.

☐   Complete the exercises:

Find the LCD of the rational expressions. Some of the denominators will need to be factored first.

1. $\dfrac{3}{10x^2}, \dfrac{7}{25x}$
2. $\dfrac{6x}{x^2-4}, \dfrac{3}{x+2}$
3. $\dfrac{2}{3t}, \dfrac{5}{t+1}$
4. $\dfrac{3}{2x^2+x}, \dfrac{2x}{6x+3}$

During Class:

☐   **Write your class notes.** Neatly write down **all** examples shown as well as key terms or phrases with definitions. If not applicable or if you were absent, watch the Lecture Series (DVD) for this section and do the same (write down the examples shown as well as key terms or phrases). Insert more paper as needed.

Class Notes/Examples	**Your Notes**

Answers: **1)** $50x^2$ **2)** $(x-2)(x+2)$ **3)** $3t(t+1)$ **4)** $3x(2x+1)$

Section 14.4 Adding and Subtracting Fractions with Different Denominators

Class Notes (continued)	**Your Notes**

(Insert additional paper as needed.)

Section 14.4 Adding and Subtracting Fractions with Different Denominators

<u>Practice</u>:

☐   Complete the Vocabulary and Readiness Check on page 1078.

☐   Next, complete any incomplete exercises below. Check and correct your work using the answers and references at the end of this section.

Review this example:

1. Add: $\dfrac{3}{10x^2}+\dfrac{7}{25x}$

Since $10x^2=2\cdot 5\cdot x\cdot x$ and $25x=5\cdot 5\cdot x$, the LCD $=50x^2$. Obtain equivalent fractions with a denominator of $50x^2$.

$$\dfrac{3}{10x^2}+\dfrac{7}{25x}=\dfrac{3(5)}{10x^2(5)}+\dfrac{7(2x)}{25x(2x)}$$

$$=\dfrac{15}{50x^2}+\dfrac{14x}{50x^2}$$

$$=\dfrac{15+14x}{50x^2}$$ Add the numerators and keep the common denominator.

Your turn:

2. Add: $\dfrac{3}{x}+\dfrac{5}{2x^2}$

Review this example:

3. Subtract: $\dfrac{7}{x-3}-\dfrac{9}{3-x}$

Notice the denominators $x-3$ and $3-x$ are opposites. That is, $-(x-3)=-x+3=3-x$. Rewrite the denominator $3-x$ as $-(x-3)$.

$$\dfrac{7}{x-3}-\dfrac{9}{3-x}$$

$$=\dfrac{7}{x-3}-\dfrac{9}{-(x-3)}$$ Recall: $\dfrac{a}{-b}=\dfrac{-a}{b}$

$$=\dfrac{7}{x-3}-\dfrac{-9}{x-3}$$

$$=\dfrac{7}{x-3}+\dfrac{9}{x-3}$$ Add the numerators and keep the common denominator.

$$=\dfrac{16}{x-3}$$

Your turn:

4. Add: $\dfrac{6}{x-3}+\dfrac{8}{3-x}$

Section 14.4 Adding and Subtracting Fractions with Different Denominators

Review this example:

5. Subtract: $\dfrac{3}{2x^2+x} - \dfrac{2x}{6x+3}$

Factor each denominator.

$$\frac{3}{2x^2+x} - \frac{2x}{6x+3} = \frac{3}{x(2x+1)} - \frac{2x}{3(2x+1)}$$

The LCD $= 3x(2x+1)$.

Obtain equivalent fractions with a denominator of $3x(2x+1)$.

$$= \frac{3(3)}{x(2x+1)(3)} - \frac{2x(x)}{3(2x+1)(x)}$$

$$= \frac{9}{3x(2x+1)} - \frac{2x^2}{3x(2x+1)}$$

$$= \boxed{\frac{9-2x^2}{3x(2x+1)}}$$ Add the numerators and keep the common denominator.

Your turn:

6. Subtract: $\dfrac{3a}{2a+6} - \dfrac{a-1}{a+3}$

Hint: Insert parentheses around binomial numerators.

$$\frac{3a}{2a+6} - \frac{(a-1)}{a+3}$$

	Answer	Text Ref	Video Ref		Answer	Text Ref	Video Ref
1	$\dfrac{15+14x}{50x^2}$	Ex 1b, p. 1074		**4**	$\dfrac{-2}{x-3}$		Sec 14.4, 2/5
2	$\dfrac{6x+5}{2x^2}$		Sec 14.4, 1/5	**5**	$\dfrac{9-2x^2}{3x(2x+1)}$	Ex 6, p. 1076	
3	$\dfrac{16}{x-3}$	Ex 4, p. 1076		**6**	$\dfrac{a+2}{2(a+3)}$		Sec 14.4, 4/5

☐   **Next, insert your homework.** Make sure you attempt all exercises asked of you and show all work, as in the exercises above. Check your answers if possible. Clearly mark any exercises you were unable to correctly complete so that you may ask questions later. DO NOT ERASE YOUR INCORRECT WORK. THIS IS HOW WE UNDERSTAND AND EXPLAIN TO YOU YOUR ERRORS.

Section 14.5 Solving Equations Containing Rational Expressions

Before Class:

☐   Read the objectives on page 1082.

☐   Read the Steps to Solve an Equation Containing Rational Expressions box on page 1084.

☐   Read the **Helpful Hint** boxes on pages 1082, 1083 and 1085.

☐   Complete the exercises:

Find the LCD of the rational expressions. Some of the denominators will need to be factored first.

1. $\dfrac{8}{5x}, \dfrac{1}{x^2}$

2. $\dfrac{5x}{x^2-9}, \dfrac{3}{x+3}$

3. $\dfrac{1}{t(t+1)}, \dfrac{5}{2(t+1)}$

During Class:

☐   **Write your class notes.** Neatly write down **all** examples shown as well as key terms or phrases with definitions. If not applicable or if you were absent, watch the Lecture Series (DVD) for this section and do the same (write down the examples shown as well as key terms or phrases). Insert more paper as needed.

Class Notes/Examples	Your Notes

Answers: **1)** $5x^2$ **2)** $(x-3)(x+3)$ **3)** $2t(t+1)$

Section 14.5 Solving Equations Containing Rational Expressions

Class Notes (continued)	**Your Notes**

(Insert additional paper as needed.)

Section 14.5 Solving Equations Containing Rational Expressions

Practice:

☐   Complete the Vocabulary and Readiness Check on page 1086.

☐   Next, complete any incomplete exercises below. Check and correct your work using the answers and references at the end of this section.

Complete this example:

1. Solve and check: $\dfrac{t-4}{2} - \dfrac{t-3}{9} = \dfrac{5}{18}$

The LCD of = $2, 9$, and 18 is 18, so we multiply both sides of the equation by 18.

$$18\left(\dfrac{t-4}{2} - \dfrac{t-3}{9}\right) = 18\left(\dfrac{5}{18}\right)$$

$18\left(\dfrac{t-4}{2}\right) - 18\left(\dfrac{t-3}{9}\right) = 18\left(\dfrac{5}{18}\right)$ Distribute.

$\qquad\quad 9(t-4) - 2(t-3) = 5$ Simplify.

$\qquad\qquad 9t - 36 - 2t + 6 = 5$ Distribute.

$\qquad\qquad\qquad\quad 7t - 30 = 5$ Combine like terms.

$\qquad\qquad\qquad\qquad\quad 7t = 35$ Solve for t.

$\qquad\qquad\qquad\qquad\quad \boxed{t = 5}$

Check: $\dfrac{5-4}{2} - \dfrac{5-3}{9} \overset{?}{=} \dfrac{5}{18}$

$\qquad\quad \dfrac{1}{2} - \dfrac{2}{9} = \dfrac{}{18} - \dfrac{}{18} =$

Your turn:

2. Solve and check: $\dfrac{x-3}{5} + \dfrac{x-2}{2} = \dfrac{1}{2}$

Review this example:

3. Solve and check: $\dfrac{2x}{x-4} = \dfrac{8}{x-4} + 1$

Multiply both sides by the LCD, $x - 4$.

$$(x-4)\left(\dfrac{2x}{x-4}\right) = (x-4)\left(\dfrac{8}{x-4} + 1\right)$$

(Solution continued on next page.)

Your turn:

4. Solve and check: $2 + \dfrac{3}{a-3} = \dfrac{a}{a-3}$

Section 14.5 Solving Equations Containing Rational Expressions

(continued)

$$(x-4) \cdot \frac{2x}{x-4} = (x-4) \cdot \frac{8}{x-4} + (x-4) \cdot 1$$
$$2x = 8 + (x-4)$$
$$2x = 4 + x$$
$$x = 4$$

Check: Notice that $x = 4$ causes the denominator to equal 0. Therefore, 4 is *not* a solution and this equation has no solution.

(continued)

Review this example:

5. Solve and check: $\dfrac{4x}{x^2 + x - 30} + \dfrac{2}{x-5} = \dfrac{1}{x+6}$

Multiply both sides by the LCD: $(x+6)(x-5)$.

$$(x+6)(x-5)\left(\frac{4x}{x^2+x-30}\right) + (x+6)(x-5)\left(\frac{2}{x-5}\right)$$
$$= (x+6)(x-5)\left(\frac{1}{x+6}\right)$$

$$4x + 2(x+6) = x - 5 \quad \text{Simplify.}$$
$$4x + 2x + 12 = x - 5 \quad \text{Distribute.}$$
$$6x + 12 = x - 5 \quad \text{Combine like terms.}$$
$$5x = -17$$
$$x = -\frac{17}{5} \quad \text{Divide both sides by 5.}$$

Check.

Your turn:

6. Solve and check:

$$\frac{4r-4}{r^2+5r-14} + \frac{2}{r+7} = \frac{1}{r-2}$$

	Answer	Text Ref	Video Ref		Answer	Text Ref	Video Ref
1	5	Ex 2, p. 1082		4	no solution		Sec 14.5, 4/6
2	3		Sec 14.5, 1/6	5	$-\dfrac{17}{5}$	Ex 4, p. 1084	
3	no solution	Ex 5, p. 1084		6	3		Sec 14.5, 5/6

☐ **Next, insert your homework.** Make sure you attempt all exercises asked of you and show all work, as in the exercises above. Check your answers if possible. Clearly mark any exercises you were unable to correctly complete so that you may ask questions later. DO NOT ERASE YOUR INCORRECT WORK. THIS IS HOW WE UNDERSTAND AND EXPLAIN TO YOU YOUR ERRORS.

Section 14.6 Rational Equations and Problem Solving

Before Class:

☐ Read the objectives on page 1091.

☐ Read the **Helpful Hint** box on page 1093.

☐ Complete the exercises:

1. It takes a painter 4 hours to paint a room in the house working alone. What part of the room can the painter complete in one hour?

2. Solve by cross multiplying: $\dfrac{3}{4} = \dfrac{1}{x}$

3. Use the equation $d = rt$ to determine the time it takes to travel 150 miles at a rate of 50 miles per hour.

4. Solve the equation $d = rt$ for time, t.

During Class:

☐ **Write your class notes.** Neatly write down **all** examples shown as well as key terms or phrases with definitions. If not applicable or if you were absent, watch the Lecture Series (DVD) for this section and do the same (write down the examples shown as well as key terms or phrases). Insert more paper as needed.

Class Notes/Examples	Your Notes

Answers: **1)** $\dfrac{1}{4}$ **2)** $x = \dfrac{4}{3} = 1\dfrac{1}{3}$ **3)** $t = 3\,\text{hrs}$ **4)** $t = \dfrac{d}{r}$

Section 14.6 Rational Equations and Problem Solving

Class Notes (continued)	Your Notes

(Insert additional paper as needed.)

Section 14.6 Rational Equations and Problem Solving

Practice:

☐   Complete the Vocabulary and Readiness Check on page 1095.

☐   Next, complete any incomplete exercises below. Check and correct your work using the answers and references at the end of this section.

Review this example:

1. Sam and Frank work in a plant that manufactures automobiles. Sam can complete a quality control tour of the plant in 3 hours while Frank takes 7 hours to complete the same job. How long will it take to complete a quality control tour if they are working together?

UNDERSTAND. Let x = the time in hours for Sam and Frank to complete the job together.

	Hours Alone	Part of Job Completed in 1 hr
Sam	3	$\dfrac{1}{3}$
Frank	7	$\dfrac{1}{7}$

TRANSLATE.

Sam's part in 1 hr		Frank's part in 1 hr		Together part in 1 hr
↓		↓		↓
$\dfrac{1}{3}$	$+$	$\dfrac{1}{7}$	$=$	$\dfrac{1}{x}$

SOLVE. Multiply both sides of the equation by $21x$.

$$21x\left(\frac{1}{3}\right) + 21x\left(\frac{1}{7}\right) = 21x\left(\frac{1}{x}\right)$$

$$7x + 3x = 21$$

$$10x = 21$$

$$x = \frac{21}{10} = 2\frac{1}{2}\text{ h working together}$$

INTERPRET. Check the answer in the original equation.

Your turn:

2. In 2 minutes a conveyor belt moves 300 pounds of recyclable aluminum from the delivery truck to a storage area. A smaller belt moves the same quantity of cans the same distance in 6 minutes. If both belts are used, find how long it takes to move the cans to the storage area.

387

Section 14.6 Rational Equations and Problem Solving

Complete this example:

3. A car travels 180 miles in the same time that a truck travels 120 miles. If the car's speed is 20 miles per hour faster than the truck's, find the car's speed and truck's speed.

UNDERSTAND. Since $d = rt$, then $\dfrac{d}{r} = t$.

Let x = the speed of the truck, then $x + 20$ = the speed of the car.

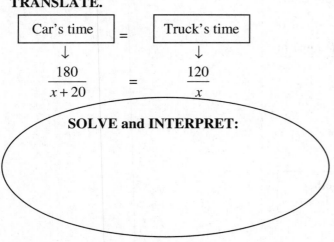

	Distance	= Rate ·	Time
Truck	120	x	$\dfrac{120}{x}$
Car	180	$x + 20$	$\dfrac{180}{x + 20}$

TRANSLATE.

$$\boxed{\text{Car's time}} = \boxed{\text{Truck's time}}$$

$$\frac{180}{x + 20} = \frac{120}{x}$$

SOLVE and INTERPRET:

Your turn:

4. A car travels 280 miles in the same time that a motorcycle travels 240 miles. If the car's speed is 10 miles per hour more than the motorcycle's, find the speed of the car and the speed of the motorcycle.

	Answer	Text Ref	Video Ref		Answer	Text Ref	Video Ref
1	$x = \dfrac{21}{10} = 2\dfrac{1}{10}$ hours	Ex 2, p. 1092		3	Car: 60 mph Truck: 40 mph Both travel for 3 hours.	Ex 3, p. 1093	
2	$1\dfrac{1}{2}$ minutes		Sec 14.6, 2/3	4	Car: 70 mph Motorcycle: 60 mph		Sec 14.6, 3/3

☐ **Next, insert your homework.** Make sure you attempt all exercises asked of you and show all work, as in the exercises above. Check your answers if possible. Clearly mark any exercises you were unable to correctly complete so that you may ask questions later. DO NOT ERASE YOUR INCORRECT WORK. THIS IS HOW WE UNDERSTAND AND EXPLAIN TO YOU YOUR ERRORS.

Section 14.7 Simplifying Complex Fractions

Before Class:

☐   Read the objectives on page 1100.

☐   Read Method 1 to Simplify a Complex Fraction on page 1100 and Method 2 to Simplify a Complex Fraction on page 1102.

☐   Read the **Helpful Hint** box on page 1102.

☐   Complete the exercises:

For questions 1 – 3, perform each operation.

1. $\dfrac{1}{3} \div \dfrac{1}{4}$ 2. $\dfrac{1}{2} - \dfrac{1}{4}$ 3. $\dfrac{1}{5} - \dfrac{1}{x}$ 4. Find the LCD for $\dfrac{1}{z}, \dfrac{1}{2}, \dfrac{1}{3}$ and $\dfrac{z}{6}$.

During Class:

☐   **Write your class notes.** Neatly write down **all** examples shown as well as key terms or phrases with definitions. If not applicable or if you were absent, watch the Lecture Series (DVD) for this section and do the same (write down the examples shown as well as key terms or phrases). Insert more paper as needed.

Class Notes/Examples	Your Notes

Answers: **1)** $\dfrac{4}{3}$ **2)** $\dfrac{1}{4}$ **3)** $\dfrac{x-5}{5x}$ **4)** $6z$

Section 14.7 Simplifying Complex Fractions

Class Notes (continued)	**Your Notes**

(Insert additional paper as needed.)

Practice:

☐   Complete the Vocabulary and Readiness Check on page 1104.

☐   Next, complete any incomplete exercises below. Check and correct your work using the answers and references at the end of this section.

Review this example:	**Your turn:**

1. Simplify: $\dfrac{\frac{5}{8}}{\frac{2}{3}}$

2. Simplify: $\dfrac{-\frac{4x}{9}}{-\frac{2x}{3}}$

Method 1: The numerator is a single fraction and the denominator is a single fraction. To divide fractions, you multiply by the reciprocal of the second fraction.

$$\frac{\frac{5}{8}}{\frac{2}{3}} = \frac{5}{8} \div \frac{2}{3} = \frac{5}{8} \cdot \frac{3}{2} = \frac{5 \cdot 3}{8 \cdot 2} = \boxed{\frac{15}{16}}$$

Review this example:	**Your turn:**

3. Simplify: $\dfrac{\frac{2}{3} + \frac{1}{5}}{\frac{2}{3} - \frac{2}{9}}$

4. Simplify: $\dfrac{\frac{1}{3}}{\frac{1}{2} - \frac{1}{4}}$

Method 1: Obtain a single fraction in the numerator and in the denominator.

$$\frac{\frac{2}{3} + \frac{1}{5}}{\frac{2}{3} - \frac{2}{9}} = \frac{\frac{2(5)}{3(5)} + \frac{1(3)}{5(3)}}{\frac{2(3)}{3(3)} - \frac{2}{9}}$$

Numerator LCD: 15

Denominator LCD: 9

$$= \frac{\frac{10}{15} + \frac{3}{15}}{\frac{6}{9} - \frac{2}{9}} = \frac{\frac{13}{15}}{\frac{4}{9}}$$

Simplify. Then, divide the fractions by multiplying by the reciprocal of the second fraction. Simplify your result.

$$= \frac{\frac{13}{15}}{\frac{4}{9}} = \frac{13}{15} \div \frac{4}{9} = \frac{13}{15} \cdot \frac{9}{4} = \frac{13 \cdot \cancel{3} \cdot 3}{\cancel{3} \cdot 5 \cdot 4} = \boxed{\frac{39}{20}}$$

Section 14.7 Simplifying Complex Fractions

Complete this example:

5. Simplify: $\dfrac{\dfrac{1}{z}-\dfrac{1}{2}}{\dfrac{1}{3}-\dfrac{z}{6}}$

Your turn:

6. Simplify: $\dfrac{\dfrac{1}{5}-\dfrac{1}{x}}{\dfrac{7}{10}+\dfrac{1}{x^2}}$

Method 1: Obtain a single fraction in the numerator and in the denominator.

$$\dfrac{\dfrac{1}{z}-\dfrac{1}{2}}{\dfrac{1}{3}-\dfrac{z}{6}}=\dfrac{\dfrac{1(2)}{z(2)}-\dfrac{1(z)}{2(z)}}{\dfrac{1(2)}{3(2)}-\dfrac{z}{6}}=\dfrac{\dfrac{2}{2z}-\dfrac{z}{2z}}{\dfrac{2}{6}-\dfrac{z}{6}}=\dfrac{\dfrac{2-z}{2z}}{\dfrac{2-z}{6}}\qquad\begin{array}{l}\text{LCD: }2z\\[4pt]\text{LCD: }6\end{array}$$

$$\dfrac{2-z}{2z}\cdot\dfrac{6}{2-z}=\dfrac{\cancel{2}\cdot 3\cdot\cancel{(2-z)}}{\cancel{2}\cdot z\cdot\cancel{(2-z)}}=\bigcirc$$

Method 2: The LCD of all the fractions is $6z$. Distribute and simplify.

$$\dfrac{\dfrac{1}{z}-\dfrac{1}{2}}{\dfrac{1}{3}-\dfrac{z}{6}}=\dfrac{6z\left(\dfrac{1}{z}-\dfrac{1}{2}\right)}{6z\left(\dfrac{1}{3}-\dfrac{z}{6}\right)}=\dfrac{6z\left(\dfrac{1}{z}\right)-6z\left(\dfrac{1}{2}\right)}{6z\left(\dfrac{1}{3}\right)-6z\left(\dfrac{z}{6}\right)}=\dfrac{6-3z}{2z-z^2}$$

Factor the numerator and denominator. Then, reduce.

$$\dfrac{6-3z}{2z-z^2}=\bigcirc\quad(\underline{\qquad\qquad})=$$

	Answer	Text Ref	Video Ref		Answer	Text Ref	Video Ref
1	$\dfrac{15}{16}$	Ex 1, p. 1100		**4**	$\dfrac{4}{3}$		Sec 14.7, 2/5
2	$\dfrac{2}{3}$		Sec 14.7, 1/5	**5**	$\dfrac{3}{z}$; $\dfrac{3(2-z)}{z(2-z)}=\dfrac{3}{z}$	Ex 3, p. 1101	
3	$\dfrac{39}{20}$	Ex 2, p. 1101		**6**	$\dfrac{2x(x-5)}{7x^2+10}$		Sec 14.7, 3–4/5

☐   **Next, insert your homework.** Make sure you attempt all exercises asked of you and show all work, as in the exercises above. Check your answers if possible. Clearly mark any exercises you were unable to correctly complete so that you may ask questions later. DO NOT ERASE YOUR INCORRECT WORK. THIS IS HOW WE UNDERSTAND AND EXPLAIN TO YOU YOUR ERRORS.

Preparing for the Chapter 14 Test

Start preparing for your Chapter 14 Test as soon as possible. Pay careful attention to any instructor discussion about this test, especially discussion on what sections you will be responsible for, etc.

☐   Work the Chapter 14 Vocabulary Check on page 1108.

☐   Read both columns (Definitions and Concepts, and Examples) of the Chapter 14 Highlights starting on page 1108.

☐   Read your Class Notes/Examples for each section covered on your Chapter 14 Test. Look for any unresolved questions you may have.

☐   Complete as many of the Chapter 14 Review exercises as possible (page 1112). Remember, the odd answers are in the back of your text.

☐   **Most important:** Place yourself in "test" conditions (see below) and work the Chapter 14 Test (pages 1116 - 1117) as a practice test the day before your actual test. To honestly assess how you are doing, try the following:

- Work on a few blank sheets of paper.
- Give yourself the same amount of time you will be given for your actual test.
- Complete this Chapter 14 Practice Test without using your notes or your text.
- If you have any time left after completing this practice test, check your work and try to find any errors on your own.
- Once done, use the back of your book to check ALL answers.
- Try to correct any errors on your own.
- Use the Chapter Test Prep Video (CTPV) to correct any errors you were unable to correct on your own. You can find these videos in the Interactive DVD Lecture Series, in MyMathLab, and on YouTube. Search Martin-Gay Prealgebra & Introductory Algebra and click "Channels."

I wish you the best of luck....Elayn Martin-Gay

Section 15.1 Introduction to Radicals

Before Class:

☐ Read the objectives on page 1122.

☐ Read the **Helpful Hint** box on page 1123.

☐ Read all italic and bold definitions below Objective A on page 1122.

☐ Complete these exercises:

1. Simplify: a. $4^2 =$ _____ b. $(-4)^2 =$ _____ c. $-4^2 =$ _____

2. Complete each list of common perfect squares.

   a. $5^2 =$ _____
      $12^2 =$ _____
      $3^2 =$ _____
      $9^2 =$ _____

   b. $6^2 =$ _____
      $1^2 =$ _____
      $10^2 =$ _____
      $8^2 =$ _____

   c. $2^2 =$ _____
      $7^2 =$ _____
      $11^2 =$ _____
      $4^2 =$ _____

During Class:

☐ **Write your class notes.** Neatly write down **all** examples shown as well as key terms or phrases with definitions. If not applicable or if you were absent, watch the Lecture Series (DVD) for this section and do the same (write down the examples shown as well as key terms or phrases). Insert more paper as needed.

Class Notes/Examples	Your Notes

Answers: **1)** a. 16 b. 16 c. −16 **2)** a. 25, 144, 9, 81 b. 36, 1, 100, 64 c. 4, 49, 121, 16

Section 15.1 Introduction to Radicals

Class Notes (continued)	Your Notes

(Insert additional paper as needed.)

Practice:

☐   Complete the Vocabulary and Readiness Check on page 1126.

☐   Next, complete any incomplete exercises below. Check and correct your work using the answers and references at the end of this section.

Review this example:	**Your turn:**
1. Find each square root.	**2.** Find each square root.
a. $-\sqrt{16}$ b. $\sqrt{\dfrac{9}{100}}$	a. $\sqrt{-4}$
a. $-\sqrt{16} = -4$	
The negative sign in front of the radical indicates the negative square root of 16.	b. $-\sqrt{121}$
b. $\sqrt{\dfrac{9}{100}} = \dfrac{3}{10}$ because $\left(\dfrac{3}{10}\right)^2 = \dfrac{9}{100}$; $\dfrac{3}{10}$ is positive.	
Review this example:	**Your turn:**
3. Find each cube root.	**4.** Find each cube root.
a. $\sqrt[3]{1}$ b. $\sqrt[3]{\dfrac{1}{125}}$	a. $\sqrt[3]{125}$
a. $\sqrt[3]{1} = 1$ because $1^3 = 1$.	b. $\sqrt[3]{-27}$
b. $\sqrt[3]{\dfrac{1}{125}} = \dfrac{1}{5}$ because $\left(\dfrac{1}{5}\right)^3 = \dfrac{1}{125}$.	
Review this example:	**Your turn:**
5. Find each root.	**6.** Find each root.
a. $\sqrt[4]{16}$ b. $\sqrt[4]{-81}$	a. $\sqrt[4]{81}$
a. $\sqrt[4]{16} = 2$ because $2^4 = 16$ and 2 is positive.	b. $-\sqrt[5]{32}$
b. $\sqrt[4]{-81}$ is not a real number since the index, 4, is even and the radicand, -81, is negative.	
Answer: Not a real number	

Section 15.1 Introduction to Radicals

Review this example:	**Your turn:**
7. Approximate $\sqrt{3}$ to three decimal places. To use a calculator, find the square root key $\sqrt{\ }$. If using a scientific calculator, press 3 then $\sqrt{\ }$. If using a graphing calculator, press $\sqrt{\ }$ then 3 then ENTER. $\boxed{\sqrt{3} \approx 1.732}$	**8.** Approximate $\sqrt{136}$ to three decimal places.

Review this example:	**Your turn:**
9. Find each root. Assume that all variables represent positive numbers. a. $\sqrt{x^6}$ b. $\sqrt[3]{-125a^{12}b^{15}}$ a. $\sqrt{x^6} = \boxed{x^3}$ because $\left(x^3\right)^2 = x^6$. b. $\sqrt[3]{-125a^{12}b^{15}} = \boxed{-5a^4b^5}$ because $\left(-5a^4b^5\right)^3 = -125a^{12}b^{15}$.	**10.** Find each root. Assume that all variables represent positive numbers. a. $\sqrt{x^4}$ b. $\sqrt{36x^{12}}$ c. $\sqrt[3]{a^6b^{18}}$ d. $\sqrt{\dfrac{x^6}{36}}$

	Answer	Text Ref	Video Ref		Answer	Text Ref	Video Ref
1	a. -4 b. $\dfrac{3}{10}$	Ex 2, 3, p. 1122		**6**	a. 3 b. -2		Sec 15.1, 12,14/21
2	a. not a real number b. -11		Sec 15.1a, 8/21	**7**	1.732	Ex 13, p. 1124	
3	a. 1 b. $\dfrac{1}{5}$	Ex 6, 8, p. 1123		**8**	11.662		Sec 15.1, 16/21
4	a. 5 b. -3		Sec 15.1, 10-11/21	**9**	a. x^3 b. $-5a^4b^5$	Ex 15, 19, p. 1125	
5	a. 2 b. not a real number	Ex 9, 12, p. 1124		**10**	a. x^2 b. $6x^6$ c. a^2b^6 d. $\dfrac{x^3}{6}$		Sec 15.1, 18-21/21

☐ **Next, insert your homework.** Make sure you attempt all exercises asked of you and show all work, as in the exercises above. Check your answers if possible. Clearly mark any exercises you were unable to correctly complete so that you may ask questions later. DO NOT ERASE YOUR INCORRECT WORK. THIS IS HOW WE UNDERSTAND AND EXPLAIN TO YOU YOUR ERRORS.

Section 15.2 Simplifying Radicals

Before Class:

☐ Read the objectives on page 1130.

☐ Read the **Helpful Hint** box on page 1130.

☐ Complete the exercises:

Find each root. Assume all variables represent positive numbers.

1. $\sqrt{16x^2}$ 2. $\sqrt{64x^6}$ 3. $\sqrt[3]{8x^3}$ 4. $\sqrt[3]{64x^6}$

During Class:

☐ **Write your class notes.** Neatly write down **all** examples shown as well as key terms or phrases with definitions. If not applicable or if you were absent, watch the Lecture Series (DVD) for this section and do the same (write down the examples shown as well as key terms or phrases). Insert more paper as needed.

Class Notes/Examples	Your Notes

Answers: **1)** $4x$ **2)** $8x^3$ **3)** $2x$ **4)** $4x^2$

Section 15.2 Simplifying Radicals

Class Notes (continued)	**Your Notes**

(Insert additional paper as needed.)

Practice:

☐ Complete the Vocabulary and Readiness Check on page 1134.

☐ Next, complete any incomplete exercises below. Check and correct your work using the answers and references at the end of this section.

Review these examples:	**Your turn:**
1. Use the product rule to simplify each radical.	**2.** Simplify. Use the product rule.

a. $\sqrt{54}$ $\qquad$ b. $\sqrt{200}$

a. $\sqrt{54}$ Factor 54 so that one of the factors is a perfect square.

$\sqrt{54} = \sqrt{9 \cdot 6}$

$\phantom{\sqrt{54}} = \sqrt{9} \cdot \sqrt{6}$ $\qquad$ Use the product rule.

$\phantom{\sqrt{54}} = \boxed{3\sqrt{6}}$ $\qquad$ Replace $\sqrt{9}$ with 3.

b. $\sqrt{200}$ Factor 200 so that one of the factors is a perfect square.

$\sqrt{200} = \sqrt{100 \cdot 2}$ $\qquad$ Use the product rule.

$\phantom{\sqrt{200}} = \sqrt{100} \cdot \sqrt{2}$

$\phantom{\sqrt{200}} = \boxed{10\sqrt{2}}$ $\qquad$ Replace $\sqrt{100}$ with 10.

Your turn:

2. Simplify. Use the product rule.

a. $\sqrt{20}$

b. $\sqrt{180}$

Review this example:	**Your turn:**
3. Simplify $3\sqrt{8}$. Use the product rule.	**4.** Simplify $-5\sqrt{27}$.

$3\sqrt{8} = 3 \cdot \sqrt{4 \cdot 2}$ $\qquad$ Factor 8 as $4 \cdot 2$.

$\phantom{3\sqrt{8}} = 3 \cdot \sqrt{4} \cdot \sqrt{2}$ $\qquad$ Use the product rule.

$\phantom{3\sqrt{8}} = 3 \cdot 2 \cdot \sqrt{2}$ $\qquad$ Replace $\sqrt{4}$ with 2.

$\phantom{3\sqrt{8}} = \boxed{6\sqrt{2}}$ $\qquad$ Replace $3 \cdot 2$ with 6.

Review this example:	**Your turn:**
5. Simplify $\sqrt{\dfrac{40}{81}}$. Use both the quotient rule and the product rule.	**6.** Simplify $\sqrt{\dfrac{27}{121}}$. Use both the quotient rule and the product rule.

$\sqrt{\dfrac{40}{81}} = \dfrac{\sqrt{40}}{\sqrt{81}} = \dfrac{\sqrt{4} \cdot \sqrt{10}}{\sqrt{81}}$

$\phantom{\sqrt{\dfrac{40}{81}}} = \boxed{\dfrac{2\sqrt{10}}{9}}$ $\qquad$ Write $\sqrt{4}$ as 2.
Write $\sqrt{81}$ as 9.

401

Section 15.2 Simplifying Radicals

Complete these examples:

7. Simplify each radical.

a. $\sqrt{x^5}$ b. $\sqrt{\dfrac{5p^3}{9}}$

Factor out the greatest even power. Then, use the product rule.

a. $\sqrt{x^5} = \sqrt{x^4 \cdot x} = \sqrt{x^4} \cdot \sqrt{x}$

$= \underline{}$ Simplify.

b. $\sqrt{\dfrac{5p^3}{9}} = \dfrac{\sqrt{5p^3}}{\sqrt{9}} = \dfrac{\sqrt{p^2 \cdot 5p}}{3} = \dfrac{\sqrt{p^2} \cdot \sqrt{5p}}{3}$

$= \underline{}$ Simplify.

Your turn:

8. Simplify each radical.

a. $\sqrt{x^{13}}$

b. $\sqrt{\dfrac{12}{m^2}}$

	Answer	Text Ref	Video Ref		Answer	Text Ref	Video Ref
1	a. $3\sqrt{6}$ b. $10\sqrt{2}$	Ex 1, 3, b. 1130 - 1131		**5**	$\dfrac{2\sqrt{10}}{9}$	Ex 8, p. 1132	
2	a. $2\sqrt{5}$ b. $6\sqrt{5}$		Sec 15.2, 1, 3/10	**6**	$\dfrac{3\sqrt{3}}{11}$		Sec 15.2, 5/10
3	$6\sqrt{2}$	Ex 5, p. 1131		**7**	a. $x^2\sqrt{x}$ b. $\dfrac{p\sqrt{5p}}{3}$	Ex 9, 12, p. 1132	
4	$-15\sqrt{3}$		Sec 15.2, 4/10	**8**	a. $x^6\sqrt{x}$ b. $\dfrac{2\sqrt{3}}{m}$		Sec 15.2, 6,8/10

☐   **Next, insert your homework.** Make sure you attempt all exercises asked of you and show all work, as in the exercises above. Check your answers if possible. Clearly mark any exercises you were unable to correctly complete so that you may ask questions later. DO NOT ERASE YOUR INCORRECT WORK. THIS IS HOW WE UNDERSTAND AND EXPLAIN TO YOU YOUR ERRORS.

Section 15.3 Adding and Subtracting Radicals

Before Class:

☐ Read the objectives on page 1138.

☐ Complete these exercises:

Fill in the blanks using the example: $\sqrt{4\cdot9} = \sqrt{4}\cdot\sqrt{9} = 2\cdot3 = 6$.

1. $\sqrt{4\cdot16} = \sqrt{\underline{}}\cdot\sqrt{\underline{}} = \underline{}\cdot\underline{} = \underline{}$

2. $\sqrt{16\cdot25} = \sqrt{\underline{}}\cdot\sqrt{\underline{}} = \underline{}\cdot\underline{} = \underline{}$

3. $\sqrt{9\cdot2} = \sqrt{\underline{}}\cdot\sqrt{\underline{}} = \underline{}\cdot\underline{} = \underline{}$

During Class:

☐ **Write your class notes.** Neatly write down **all** examples shown as well as key terms or phrases with definitions. If not applicable or if you were absent, watch the Lecture Series (DVD) for this section and do the same (write down the examples shown as well as key terms or phrases). Insert more paper as needed.

Class Notes/Examples	Your Notes

Answers: **1)** $\sqrt{4\cdot16} = \sqrt{4}\cdot\sqrt{16} = 2\cdot4 = 8$ **2)** $\sqrt{16\cdot25} = \sqrt{16}\cdot\sqrt{25} = 4\cdot5 = 20$
3) $\sqrt{9\cdot2} = \sqrt{9}\cdot\sqrt{2} = 3\cdot\sqrt{2} = 3\sqrt{2}$

Section 15.3 Adding and Subtracting Radicals

Class Notes (continued)	**Your Notes**

(Insert additional paper as needed.)

Section 15.3 Adding and Subtracting Radicals

Practice:

☐   Complete the Vocabulary and Readiness Check on page 1140.

☐   Next, complete any incomplete exercises below. Check and correct your work using the answers and references at the end of this section.

Review this example:
1. Add or subtract as indicated.

   a. $4\sqrt{5}+3\sqrt{5}$ b. $\sqrt{10}-6\sqrt{10}$

   c. $2\sqrt{6}+2\sqrt{5}$

a. $4\sqrt{5}+3\sqrt{5}=(4+3)\sqrt{5}=\boxed{7\sqrt{5}}$

b. $\sqrt{10}-6\sqrt{10}=1\sqrt{10}-6\sqrt{10}$
$\phantom{\sqrt{10}-6\sqrt{10}}=(1-6)\sqrt{10}=\boxed{-5\sqrt{10}}$

c. $2\sqrt{6}+2\sqrt{5}$ is (already simplified.)
   The radicands are not the same.

Your turn:
2. Add or subtract as indicated.

   a. $4\sqrt{3}-8\sqrt{3}$

   b. $3\sqrt{6}+8\sqrt{6}-2\sqrt{6}-5$

Review this example:
3. Simplify the radical expression.

   $\sqrt{50}+\sqrt{8}$

$\begin{aligned}\sqrt{50}+\sqrt{8} &= \sqrt{25\cdot2}+\sqrt{4\cdot2} &&\text{Factor radicands.}\\ &= \sqrt{25}\sqrt{2}+\sqrt{4}\sqrt{2} &&\text{Use the product rule.}\\ &= 5\sqrt{2}+2\sqrt{2} &&\text{Simplify.}\\ &= \boxed{7\sqrt{2}} &&\text{Add like radicals.}\end{aligned}$

Your turn:
4. Simplify the radical expression.

   $\sqrt{12}+\sqrt{27}$

Complete this example:
5. Simplify $7\sqrt{12}-2\sqrt{75}$.

$\begin{aligned}&7\sqrt{12}-2\sqrt{75}\\ &=7\sqrt{4\cdot3}-2\sqrt{25\cdot3} &&\text{Factor radicands.}\\ &=7\sqrt{4}\sqrt{3}-2\sqrt{25}\sqrt{3} &&\text{Use the product rule.}\\ &=7\cdot2\sqrt{3}-2\cdot5\sqrt{3} &&\text{Simplify.}\\ &=14\sqrt{3}-10\sqrt{3} &&\text{Multiply.}\\ &=\boxed{} &&\text{Subtract like radicals.}\end{aligned}$

Your turn:
6. Simplify the radical expression.

   $\sqrt{\dfrac{3}{64}}+\sqrt{\dfrac{3}{16}}$

Section 15.3 Adding and Subtracting Radicals

Review this example:

7. Simplify: $2\sqrt{x^2} - \sqrt{25x^5} + \sqrt{x^5}$

Hint: $\sqrt{x^5}$ factors as $\sqrt{x^4 \cdot x}$, so one factor is a perfect square.

$= 2x - \sqrt{25x^4 \cdot x} + \sqrt{x^4 \cdot x}$ Factor radicands.

$= 2x - \sqrt{25x^4}\sqrt{x} + \sqrt{x^4}\sqrt{x}$ Use the product rule.

$= 2x - 5x^2\sqrt{x} + x^2\sqrt{x}$ Simplify.

$= 2x - 4x^2\sqrt{x}$ Add like radicals.

Your turn:

8. Simplify: $5\sqrt{2x} + \sqrt{98x}$

Review this example:

9. Simplify: $5\sqrt[3]{16x^3} - \sqrt[3]{54x^3}$

Factor radicands so that one factor is a perfect cube.

$= 5\sqrt[3]{8x^3 \cdot 2} - \sqrt[3]{27x^3 \cdot 2}$ Factor radicands.

$= 5 \cdot \sqrt[3]{8x^3} \cdot \sqrt[3]{2} - \sqrt[3]{27x^3} \cdot \sqrt[3]{2}$ Use the product rule.

$= 5 \cdot 2x \cdot \sqrt[3]{2} - 3x \cdot \sqrt[3]{2}$ Simplify.

$= 10x \cdot \sqrt[3]{2} - 3x \cdot \sqrt[3]{2}$ Multiply.

$= 7x\sqrt[3]{2}$ Subtract like radicals.

Your turn:

10. Simplify: $\sqrt[3]{8} + \sqrt[3]{54} - 5$

	Answer	Text Ref	Video Ref		Answer	Text Ref	Video Ref
1	a. $7\sqrt{5}$ b. $-5\sqrt{10}$ c. already simplified	Ex 1 - 3, p. 1138		6	$\dfrac{3\sqrt{3}}{8}$		Sec 15.3, 7/8
2	a. $-4\sqrt{3}$ b. $9\sqrt{6} - 5$		Sec 15.3, 2, 1/8	7	$2x - 4x^2\sqrt{x}$	Ex 8, p. 1139	
3	$7\sqrt{2}$	Ex 5, p. 1139		8	$12\sqrt{2x}$		Sec 15.3, 6/8
4	$5\sqrt{3}$		Sec 15.3, 5/8	9	$7x\sqrt[3]{2}$	Ex 9, p. 1139	
5	$4\sqrt{3}$	Ex 6, p. 1139		10	$-3 + 3\sqrt[3]{2}$		Sec 15.3, 8/8

☐   **Next, insert your homework.** Make sure you attempt all exercises asked of you and show all work, as in the exercises above. Check your answers if possible. Clearly mark any exercises you were unable to correctly complete so that you may ask questions later. DO NOT ERASE YOUR INCORRECT WORK. THIS IS HOW WE UNDERSTAND AND EXPLAIN TO YOU YOUR ERRORS.

Section 15.4 Multiplying And Dividing Radicals

Before Class:

☐ Read the objectives on page 1143.

☐ Read the Helpful Hint box on page 1146.

☐ Complete the exercises:

Simplify each radical. Assume all variables represent positive numbers.

1. $\sqrt{24}$ 2. $\sqrt{x^7}$ 3. $\sqrt{18x^3}$ 4. $\sqrt{45x^2}$

During Class:

☐ **Write your class notes.** Neatly write down **all** examples shown as well as key terms or phrases with definitions. If not applicable or if you were absent, watch the Lecture Series (DVD) for this section and do the same (write down the examples shown as well as key terms or phrases). Insert more paper as needed.

Class Notes/Examples	**Your Notes**

Answers: **1)** $2\sqrt{6}$ **2)** $x^3\sqrt{x}$ **3)** $3x\sqrt{2x}$ **4)** $3x\sqrt{5}$

407

Section 15.4 Multiplying And Dividing Radicals

Class Notes (continued)

Your Notes

(Insert additional paper as needed.)

Section 15.4 Multiplying And Dividing Radicals

Practice:

☐   Complete the Vocabulary and Readiness Check on page 1148.

☐   Next, complete any incomplete exercises below. Check and correct your work using the
        answers and references at the end of this section.

Review this example:	**Your turn:**
1. Multiply. Simplify if possible.	**2.** Multiply. Simplify if possible.

$\sqrt{3} \cdot \sqrt{15}$ | $\sqrt{10} \cdot \sqrt{5}$

$\sqrt{3} \cdot \sqrt{15} = \sqrt{45}$ Use the product rule.

$\qquad = \sqrt{9 \cdot 5}$ Factor 45 as $9 \cdot 5$.

$\qquad = \sqrt{9} \cdot \sqrt{5}$ Use the product rule.

$\qquad = 3\sqrt{5}$ Replace $\sqrt{9}$ with 3.

Review this example:

3. Multiply. Then, simplify.

$\sqrt{5}(\sqrt{5} - \sqrt{2})$

Distribute and simplify.

$\sqrt{5}(\sqrt{5} - \sqrt{2}) = \sqrt{5} \cdot \sqrt{5} - \sqrt{5} \cdot \sqrt{2}$

$\qquad\qquad = \sqrt{25} - \sqrt{10}$

$\qquad\qquad = 5 - \sqrt{10}$

Your turn:

4. Multiply. Then, simplify.

$\sqrt{6}(\sqrt{5} + \sqrt{7})$

Review this example:

5. Divide $\dfrac{\sqrt{12x^3}}{\sqrt{3x}}$. Then, simplify.

Use the quotient rule. Then, simplify.

$\dfrac{\sqrt{12x^3}}{\sqrt{3x}} = \sqrt{\dfrac{12x^3}{3x}} = \sqrt{4x^2} = 2x$

Your turn:

6. Divide $\dfrac{\sqrt{75y^5}}{\sqrt{3y}}$. Then, simplify.

409

Section 15.4 Multiplying And Dividing Radicals

Review this example:

7. Rationalize the denominator.

$$\frac{2}{1+\sqrt{3}}$$

Multiply the numerator and the denominator of the fraction by the conjugate of the denominator.

$$\frac{2}{1+\sqrt{3}} = \frac{2(1-\sqrt{3})}{(1+\sqrt{3})(1-\sqrt{3})}$$

$$= \frac{2(1-\sqrt{3})}{(1)^2 - (\sqrt{3})^2}$$

$$= \frac{2(1-\sqrt{3})}{1-3}$$

$$= \frac{2(1-\sqrt{3})}{-2}$$

$$= -1(1-\sqrt{3})$$

$$= \boxed{-1+\sqrt{3}}$$

The conjugate of $1+\sqrt{3}$ is $1-\sqrt{3}$.

For conjugates:
$$(a-b)(a+b) = a^2 - b^2$$

Replace: $(1)^2 = 1$
$\qquad\qquad (\sqrt{3})^2 = 3$

Divide common factors:
$$\frac{2}{-2} = -1.$$
Then, distribute.

Your turn:

8. Rationalize the denominator.

$$\frac{4}{2-\sqrt{5}}$$

	Answer	Text Ref	Video Ref		Answer	Text Ref	Video Ref
1	$3\sqrt{5}$	Ex 3, p. 1143		5	$2x$	Ex 9, p. 1145	
2	$5\sqrt{2}$		Sec 15.4, 1/9	6	$5y^2$		Sec 15.4, 6/9
3	$5-\sqrt{10}$	Ex 5a, p. 1144		7	$-1+\sqrt{3}$	Ex 13, p. 1146	
4	$\sqrt{30}+\sqrt{42}$		Sec 15.4, 3/9	8	$-8-4\sqrt{5}$		Sec 15.4, 9/9

☐ **Next, insert your homework.** Make sure you attempt all exercises asked of you and show all work, as in the exercises above. Check your answers if possible. Clearly mark any exercises you were unable to correctly complete so that you may ask questions later. DO NOT ERASE YOUR INCORRECT WORK. THIS IS HOW WE UNDERSTAND AND EXPLAIN TO YOU YOUR ERRORS.

Section 15.5 Solving Equations Containing Radicals

Before Class:

☐   Read the objectives on page 1153.

☐   Read the **Helpful Hint** boxes on pages 1153, 1155 and 1156.

☐   Read the To Solve a Radical Equation Containing Square Roots box on page 1155.

☐   Complete these exercises:

Identify each statement as true or false.

1. $\sqrt{6} \cdot \sqrt{6} = 6$ _____

2. $\sqrt{9} \cdot \sqrt{3} = 3\sqrt{3}$ _____

3. $\sqrt{5x} \cdot \sqrt{5x} = 2\sqrt{5x}$ _____

4. $\sqrt{5} + \sqrt{5} = \sqrt{10}$ _____

5. $\sqrt{2} \cdot \sqrt{8} = 4$ _____

6. $\sqrt{7x} + \sqrt{7x} = 2\sqrt{7x}$ _____

During Class:

☐   **Write your class notes.** Neatly write down **all** examples shown as well as key terms or phrases with definitions. If not applicable or if you were absent, watch the Lecture Series (DVD) for this section and do the same (write down the examples shown as well as key terms or phrases). Insert more paper as needed.

Class Notes/Examples	**Your Notes**

Answers: **1)** True   **2)** True   **3)** False   **4)** False   **5)** True   **6)** True

Section 15.5 Solving Equations Containing Radicals

Class Notes (continued)	**Your Notes**

(Insert additional paper as needed.)

Section 15.5 Solving Equations Containing Radicals

Practice:

☐   Complete any incomplete exercises below. Check and correct your work using the answers and references at the end of this section.

Complete this example:

1. Solve: $\sqrt{x+3} = 5$

$(\sqrt{x+3})^2 = (5)^2$ Square both sides.

$x + 3 = 25$ Simplify.

$\boxed{x = 22}$ Subtract 3 from both sides.

Check: $\sqrt{22+3} = \underline{\quad} = \underline{\quad}$

Your turn:

2. Solve and check: $\sqrt{x+5} = 2$

Review this example:

3. Solve: $\sqrt{x} + 6 = 4$

$\sqrt{x} = -2$ Isolate the radical. Subtract 6.

$(\sqrt{x})^2 = (-2)^2$ Square both sides.

$x = 4$ Simplify.

Check: $\sqrt{4} + 6 = 2 + 6 = 8 \neq 4$
This equation has $\boxed{\text{no solution.}}$
Recall $\sqrt{x}$ is the nonnegative square root of x.

Your turn:

4. Solve and check: $3\sqrt{x} + 5 = 2$

Complete this example:

5. Solve: $\sqrt{x-4} = \sqrt{x} - 2$

$(\sqrt{x-4})^2 = (\sqrt{x}-2)^2$ Square both sides.

$x - 4 = (\sqrt{x}-2)(\sqrt{x}-2)$ Simplify.

$x - 4 = \sqrt{x} \cdot \sqrt{x} - 2\sqrt{x} - 2\sqrt{x} + 4$ Distribute twice.

$x - 4 = x - 4\sqrt{x} + 4$ Simplify.

$-8 = -4\sqrt{x}$ Subtract x, subtract 4.

$2 = \sqrt{x}$ Divide both sides by –4.

$\boxed{4 = x}$ Square both sides again.

Check:

Your turn:

6. Solve and check:
$\sqrt{x-7} = \sqrt{x} - 1$

Section 15.5 Solving Equations Containing Radicals

Complete this example:

7. Solve: $\sqrt{x+3} - x = -3$

$\sqrt{x+3} = x-3$	Isolate radical, add x.
$(\sqrt{x+3})^2 = (x-3)^2$	Square both sides.
$x+3 = (x-3)(x-3)$	Simplify.
$x+3 = x^2 - 3x - 3x + 9$	Distribute twice.
$x+3 = x^2 - 6x + 9$	Simplify.
$3 = x^2 - 7x + 9$	Subtract x from both sides.
$0 = x^2 - 7x + 6$	Subtract 3 from both sides.
$0 = (x-6)(x-1)$	Factor.
$0 = (x-6)$ or $0 = (x-1)$	Set each factor = 0.
$6 = x \qquad\qquad 1 = x$	

Check both answers.

Let $x = 6$.   Let $x = 1$.

$\underline{\qquad\qquad}\overset{?}{=} -3$ $\underline{\qquad\qquad}\overset{?}{=} -3$

The solution is (_____).

Your turn:

8. Solve and check: $\sqrt{1-8x} - x = 4$

	Answer	Text Ref	Video Ref		Answer	Text Ref	Video Ref
1	22	Ex 1, p. 1153		5	4	Ex 6, p. 1156	
2	−1		Sec 15.5, 1/5	6	16		Sec 15.5, 5/5
3	No solution	Ex 3, p. 1154		7	6	Ex 5, p. 1155	
4	No solution		Sec 15.5, 2/5	8	−1		Sec 15.5, 4/5

☐   **Next, insert your homework.** Make sure you attempt all exercises asked of you and show all work, as in the exercises above. Check your answers if possible. Clearly mark any exercises you were unable to correctly complete so that you may ask questions later. DO NOT ERASE YOUR INCORRECT WORK. THIS IS HOW WE UNDERSTAND AND EXPLAIN TO YOU YOUR ERRORS.

Section 15.6 Radical Equations And Problem Solving

Before Class:

☐   Read the objectives on page 1159.

☐   Read the Pythagorean Theorem box on page 1159.

☐   Complete the exercises:

Solve for the variable in each equation.   The variable in each equation represents a length. Give an exact answer.

1. $36 + 64 = c^2$ 2. $4 + b^2 = 25$ 3. $a^2 + 9 = 36$

During Class:

☐   **Write your class notes.**   Neatly write down **all** examples shown as well as key terms or phrases with definitions. If not applicable or if you were absent, watch the Lecture Series (DVD) for this section and do the same (write down the examples shown as well as key terms or phrases). Insert more paper as needed.

Class Notes/Examples	Your Notes

Answers: **1)** $c = 10$ **2)** $b = \sqrt{21}$ **3)** $a = 3\sqrt{3}$

Section 15.6 Radical Equations And Problem Solving

Class Notes (continued)	**Your Notes**

(Insert additional paper as needed.)

Section 15.6 Radical Equations And Problem Solving

Practice:

☐   Complete any incomplete exercises below. Check and correct your work using the answers and references at the end of this section.

Review this example:

1. Use the Pythagorean theorem to find the length of the unknown side of the right triangle. Give an exact answer and a two-decimal-place approximation.

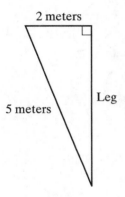

2 meters

Leg

5 meters

Let $a = 2$ meters and b = the length of the other leg. The hypotenuse is $c = 5$ meters.

$a^2 + b^2 = c^2$	Use the Pythagorean theorem.
$2^2 + b^2 = 5^2$	Let $a = 2$ and $c = 5$.
$4 + b^2 = 25$	Solve for b.
$b^2 = 21$	

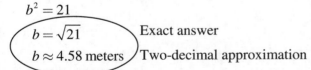

$b = \sqrt{21}$ Exact answer

$b \approx 4.58$ meters   Two-decimal approximation

Your turn:

2. Use the Pythagorean theorem to find the length of the unknown side of the right triangle. Give an exact answer and a two-decimal-place approximation.

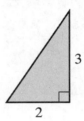

3

2

Section 15.6 Radical Equations And Problem Solving

Review this example:

3. A surveyor must determine the distance across a lake at points P and Q as shown in the figure. To do this, she finds a third point, R, perpendicular to line PQ. If the length of $\overline{PR}$ is 320 feet and the length of $\overline{QR}$ is 240 feet, what is the distance across the lake? Approximate this distance to the nearest whole foot.

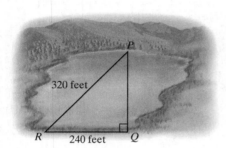

Your turn:

4. A wire is used to anchor a 20-foot pole. One end of the wire is attached to the top of the pole. The other end is fastened to a stake five feet away from the bottom of the pole. Find the length of the wire rounded to the nearest tenth of a foot.

UNDERSTAND and TRANSLATE.
Let $c = 320$, $a = 240$ and $b =$ the unknown length.

$a^2 + b^2 = c^2$ Use the Pythagorean theorem.
$240^2 + b^2 = 320^2$ Substitute $c = 320$ and $a = 240$.

SOLVE and INTERPRET.
$57,600 + b^2 = 102,400$

$b^2 = 44,800$ Subtract 57,600 from both sides.

$b = \sqrt{44,800} = \sqrt{6400} \cdot \sqrt{7} = 80\sqrt{7} \approx 212$ ft

The distance across the lake is ≈ 212 ft.

	Answer	Text Ref	Video Ref		Answer	Text Ref	Video Ref
1	exactly $\sqrt{21}$ m, approx. 4.58 m	Ex 2, p. 1159		3	212 ft	Ex 3, p. 1160	
2	exactly $\sqrt{13}$, approx. 3.61		Sec 15.6, 1/4	4	20.6 ft		Sec 15.6, 3/4

☐   **Next, insert your homework.** Make sure you attempt all exercises asked of you and show all work, as in the exercises above. Check your answers if possible. Clearly mark any exercises you were unable to correctly complete so that you may ask questions later. DO NOT ERASE YOUR INCORRECT WORK. THIS IS HOW WE UNDERSTAND AND EXPLAIN TO YOU YOUR ERRORS.

Preparing for the Chapter 15 Test

Start preparing for your Chapter 15 Test as soon as possible. Pay careful attention to any instructor discussion about this test, especially discussion on what sections you will be responsible for, etc.

☐ Work the Chapter 15 Vocabulary Check on page 1166.

☐ Read both columns (Definitions and Concepts, and Examples) of the Chapter 15 Highlights starting on page 1166.

☐ Read your Class Notes/Examples for each section covered on your Chapter 15 Test. Look for any unresolved questions you may have.

☐ Complete as many of the Chapter 15 Review exercises as possible (starting on page 1169). Remember, the odd answers are in the back of your text.

☐ **Most important:** Place yourself in "test" conditions (see below) and work the Chapter 15 Test (pages 1172 - 1173) as a practice test the day before your actual test. To honestly assess how you are doing, try the following:
- Work on a few blank sheets of paper.
- Give yourself the same amount of time you will be given for your actual test.
- Complete this Chapter 15 Practice Test without using your notes or your text.
- If you have any time left after completing this practice test, check your work and try to find any errors on your own.
- Once done, use the back of your book to check ALL answers.
- Try to correct any errors on your own.
- Use the Chapter Test Prep Video (CTPV) to correct any errors you were unable to correct on your own. You can find these videos in the Interactive DVD Lecture Series, in MyMathLab, and on YouTube. Search MartinGayDevMath and click "Channels."

I wish you the best of luck....Elayn Martin-Gay

Section 16.1 Solving Quadratic Equations by the Square Root Property

Before Class:

☐   Read the objectives on page 1178.

☐   Read the **Helpful Hint** box on page 1180.

☐   Complete these exercises:

1.   Simplify, if possible:

a. $\sqrt{12} = \underline{\hspace{1cm}}$ b. $\sqrt{32} = \underline{\hspace{1cm}}$ c. $\sqrt{18} = \underline{\hspace{1cm}}$ d. $\sqrt{10} = \underline{\hspace{1cm}}$

2.   Solve by factoring: $x^2 - 9 = 0$

3.   Solve by factoring: $x^2 + 6x - 7 = 0$

During Class:

☐   **Write your class notes.** Neatly write down **all** examples shown as well as key terms or phrases with definitions. If not applicable or if you were absent, watch the Lecture Series (DVD) for this section and do the same (write down the examples shown as well as key terms or phrases). Insert more paper as needed.

Class Notes/Examples	**Your Notes**

Answers: **1)**   a. $2\sqrt{3}$ b. $4\sqrt{2}$ c. $3\sqrt{2}$ b. $\sqrt{10}$ **2)** $x = 3, -3$ **3)** $x = 1, -7$

Section 16.1 Solving Quadratic Equations by the Square Root Property

Class Notes (continued)	**Your Notes**

(Insert additional paper as needed.)

Section 16.1 Solving Quadratic Equations by the Square Root Property

Practice:

☐   Complete any incomplete exercises below. Check and correct your work using the answers and references at the end of this section.

Complete this example:	**Your turn:**
1. Use the square root property to solve: $x^2 - 9 = 0$	**2.** Use the square root property to solve: $x^2 = 64$

$$x^2 - 9 = 0 \qquad \text{Add 9 to both sides.}$$
$$x^2 = 9 \qquad \text{Use the square root property.}$$
$$\sqrt{x^2} = \sqrt{9}$$
$$x = \sqrt{9} \quad \text{or} \quad x = -\sqrt{9}$$
$$x = 3 \qquad\qquad x = -3$$

Check: $3^2 - 9 \overset{?}{=} 0 \qquad (-3)^2 - 9 \overset{?}{=} 0$

—————   —————

Solutions are   and   .

Review this example:	**Your turn:**
3. Use the square root property to solve: $(x-1)^2 = -2$	**4.** Use the square root property to solve: $x^2 = -4$

This equation has ⟨no real solution⟩ because the square root of -2 is not a real number.

Complete this example:	**Your turn:**
5. Solve: $(x+1)^2 = 8$	**6.** Use the square root property to solve: $(p+2)^2 = 10$

Use the square root property.

$$x + 1 = \sqrt{8} \qquad\qquad x + 1 = -\sqrt{8}$$
$$x + 1 = 2\sqrt{2} \quad \text{or} \quad x + 1 = -2\sqrt{2} \qquad \text{Simplify } \sqrt{8}.$$
$$x = -1 + 2\sqrt{2} \qquad x = -1 - 2\sqrt{2} \qquad \text{Solve for } x.$$

These can be written compactly as $-1 \pm 2\sqrt{2}$

Check both :

$$(-1 + 2\sqrt{2} + 1)^2 \overset{?}{=} 8 \qquad (-1 - 2\sqrt{2} + 1)^2 \overset{?}{=} 8$$

423

Section 16.1 Solving Quadratic Equations by the Square Root Property

Complete this example:

7. Use the square root property to
solve: $(5x-2)^2 = 10$

Use the square root property.

$5x-2 = \sqrt{10}$ or $5x-2 = -\sqrt{10}$

Add 2 to both sides.

$5x = 2 + \sqrt{10}$ or $5x = 2 - \sqrt{10}$

Divide both sides by 5.

$x = \dfrac{2+\sqrt{10}}{5}$ or $x = \dfrac{2-\sqrt{10}}{5}$

These can be written compactly as $\boxed{\dfrac{2\pm\sqrt{10}}{5}}$.

Check both:

$\left[5\left(\dfrac{2+\sqrt{10}}{5}\right) - 2\right]^2 \overset{?}{=} 10$ $\left[5\left(\dfrac{2-\sqrt{10}}{5}\right) - 2\right]^2 \overset{?}{=} 10$

Your turn:

8. Use the square root property to
solve: $(3x-7)^2 = 32$

	Answer	Text Ref	Video Ref		Answer	Text Ref	Video Ref
1	$3, -3$	Ex 3, p. 1179		**5**	$-1 \pm 2\sqrt{2}$	Ex 6, p. 1180	
2	± 8		Sec 16.1, 1/6	**6**	$-2 \pm \sqrt{10}$		Sec 16.1, 4/6
3	No real solution	Ex 7, p. 1180		**7**	$\dfrac{2 \pm \sqrt{10}}{5}$	Ex 8, p. 1180	
4	No real solution		Sec 16.1, 2/6	**8**	$\dfrac{7 \pm 4\sqrt{2}}{3}$		Sec 16.1, 5/6

☐   **Next, insert your homework.** Make sure you attempt all exercises asked of you and show all work, as in the exercises above. Check your answers if possible. Clearly mark any exercises you were unable to correctly complete so that you may ask questions later. DO NOT ERASE YOUR INCORRECT WORK. THIS IS HOW WE UNDERSTAND AND EXPLAIN TO YOU YOUR ERRORS.

Section 16.2 Solving Quadratic Equations by Completing the Square

Before Class:

☐   Read the objectives on page 1184.

☐   Read the Completing the Square box on page 1185.

☐   Read the **Helpful Hint** box on page 1185.

☐   Complete the exercises: Solve each quadratic equation using the square root property.

     1. $(x+1)^2 = 7$ 2. $(x+1)^2 = 4$ 3. $(x+1)^2 = 8$

During Class:

☐   **Write your class notes.** Neatly write down **all** examples shown as well as key terms or phrases with definitions. If not applicable or if you were absent, watch the Lecture Series (DVD/CD) for this section and do the same (write down the examples shown as well as key terms or phrases). Insert more paper as needed.

Class Notes/Examples	**Your Notes**

Answers: **1)** $x = -1 \pm \sqrt{7}$ **2)** $x = 1, x = -3$ **3)** $x = -1 \pm 2\sqrt{2}$

Section 16.2 Solving Quadratic Equations by Completing the Square

Class Notes (continued)	**Your Notes**

(Insert additional paper as needed.)

Section 16.2 Solving Quadratic Equations by Completing the Square

Practice:

☐   Complete the Vocabulary and Readiness Check on page 1187.

☐   Next, complete any incomplete exercises below. Check and correct your work using the answers and references at the end of this section.

Review this example:	**Your turn:**

1. Solve by completing the square.

$$x^2 + 6x + 3 = 0$$

Get the variable terms alone on the left side of the equation.

$x^2 + 6x + 3 = 0$
$\quad x^2 + 6x = -3$ Subtract 3 from both sides.

Next, we will find half of the coefficient of the x-term, and then square it. This number will be added to both sides of the equation.

$x^2 + 6x + \underline{\quad} = -3 + \underline{\quad}$

Half of 6 is 3, and 3 squared is 9.

$x^2 + 6x + \underline{9} = -3 + \underline{9}$ Complete the square by adding 9 to both sides.

$\quad x^2 + 6x + 9 = -3 + 9$ Simplify.
$\quad x^2 + 6x + 9 = 6$ Factor.
$\quad (x+3)(x+3) = 6$ Write as a square.
$\quad (x+3)^2 = 6$

Use the square root property.

$x + 3 = \sqrt{6}$ or $x + 3 = -\sqrt{6}$
$\quad x = -3 + \sqrt{6}$ $\quad x = -3 - \sqrt{6}$

The solutions are $x = -3 \pm \sqrt{6}$.

Your turn:

2. Solve by completing the square.

$$x^2 + 8x = -12$$

Section 16.2 Solving Quadratic Equations by Completing the Square

Complete this example:

3. Solve the quadratic equation by completing the square.

$$x^2 - 10x = -14$$

$x^2 - 10x = -14$ Variable terms are alone.

Half of -10 is -5, and $(-5)^2$ squared is 25.

$x^2 - 10x + \underline{25} = -14 + \underline{25}$ Complete the square by adding 25 to both sides.

$x^2 - 10x + 25 = -14 + 25$

$x^2 - 10x + 25 = 11$ Simplify.

$(x-5)(x-5) = 11$ Factor.

$(x-5)^2 = 11$

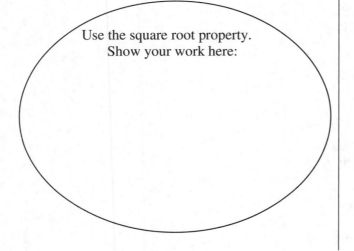

Use the square root property.
Show your work here:

Your turn:

4. Solve the quadratic equation by completing the square.

$$x^2 - 2x - 1 = 0$$

	Answer	Text Ref	Video Ref		Answer	Text Ref	Video Ref
1	$x = -3 \pm \sqrt{6}$	Ex 1, p 1185		3	$x = 5 \pm \sqrt{11}$	Ex 2, p 1185	
2	$x = -6, x = -2$		Sec 16.2, 3/5	4	$x = 1 \pm \sqrt{2}$		Sec 16.2, 4/5

☐ **Next, insert your homework.** Make sure you attempt all exercises asked of you and show all work, as in the exercises above. Check your answers if possible. Clearly mark any exercises you were unable to correctly complete so that you may ask questions later. DO NOT ERASE YOUR INCORRECT WORK. THIS IS HOW WE UNDERSTAND AND EXPLAIN TO YOU YOUR ERRORS.

Section 16.3 Solving Quadratic Equations by the Quadratic Formula

Before Class:

☐   Read the objectives on page 1190.

☐   Read the **Helpful Hint** boxes on pages 1191, 1192, and 1194.

☐   Complete these exercises:

Fill in the blank with the number needed to make each expression a perfect square trinomial.

1. $x^2 + 6x + \underline{\quad}$ 2. $y^2 + 10y + \underline{\quad}$

Evaluate $\sqrt{b^2 - 4ac}$ given

3. $a = 2, b = -9,$ and $c = -5$.

4. $a = 2, b = 6,$ and $c = -3$

During Class:

☐   **Write your class notes.** Neatly write down **all** examples shown as well as key terms or phrases with definitions. If not applicable or if you were absent, watch the Lecture Series (DVD) for this section and do the same (write down the examples shown as well as key terms or phrases). Insert more paper as needed.

Class Notes/Examples	**Your Notes**

Answers: **1)** 9   **2)** 25   **3)** $\sqrt{121} = 11$ **4)** $\sqrt{60} = 2\sqrt{15}$

Section 16.3 Solving Quadratic Equations by the Quadratic Formula

Class Notes (continued)	**Your Notes**

(Insert additional paper as needed.)

Section 16.3 Solving Quadratic Equations by the Quadratic Formula

Practice:

☐   Complete the Vocabulary and Readiness Check on page 1194.

☐   Complete any incomplete exercises below. Check and correct your work using the answers and references at the end of this section.

Review this example:

1. Solve $3x^2 + x - 3 = 0$ using the quadratic formula.

This equation is in standard form $ax^2 + bx + c = 0$ with $a = 3, b = 1,$ and $c = -3$.

Use the quadratic formula.

$$x = \frac{-b \pm \sqrt{b^2 - 4ac}}{2a}$$

$$x = \frac{-1 \pm \sqrt{1^2 - 4 \cdot 3 \cdot (-3)}}{2 \cdot 3} \quad \text{Let } a = 3, b = 1, c = -3.$$

$$x = \frac{-1 \pm \sqrt{1 + 36}}{6} \quad \text{Simplify.}$$

$$= \frac{-1 \pm \sqrt{37}}{6}$$

Your turn:

2. Solve $3k^2 + 7k + 1 = 0$ using the quadratic formula.

Review this example:

3. Solve $2x^2 - 9x = 5$ using the quadratic formula.

$2x^2 - 9x - 5 = 0$ Rewritten in standard form.
                    $a = 2, b = -3, c = -5$

Substitute into the quadratic formula (above).

$$x = \frac{-(-9) \pm \sqrt{(-9)^2 - 4 \cdot 2 \cdot (-5)}}{2 \cdot 2}$$

$$x = \frac{9 \pm \sqrt{81 + 40}}{4}$$

$$x = \frac{9 \pm \sqrt{121}}{4} = \frac{9 \pm 11}{4}$$

$$x = \frac{9 - 11}{4} = -\frac{2}{4} = -\frac{1}{2} \quad \text{or} \quad x = \frac{9 + 11}{4} = \frac{20}{4} = 5$$

Your turn:

4. Solve $3 - x^2 = 4x$ using the quadratic formula.

431

Section 16.3 Solving Quadratic Equations by the Quadratic Formula

Review this example:

5. Solve $\frac{1}{2}x^2 - x = 2$ using the quadratic formula.

$\frac{1}{2}x^2 - x = 2$ Rewritten in standard form.

Clear the equation of fractions.

$2\left(\frac{1}{2}x^2 - x\right) = 2 \cdot 2$ Multiply by the LCD, 2.

$\quad\quad x^2 - x = 4$ Subtract 4 from both sides.

$x^2 - x - 4 = 0$ Here $a = 1$, $b = -2$, $c = -4$.

Substitute into the quadratic formula.

$x = \dfrac{-(-2) \pm \sqrt{(-2)^2 - 4 \cdot 1 \cdot (-4)}}{2 \cdot 1}$

$x = \dfrac{2 \pm \sqrt{4 + 16}}{2} = \dfrac{2 \pm \sqrt{20}}{2} = \dfrac{2 \pm 2\sqrt{5}}{2}$ Simplify.

$x = \dfrac{2(1 \pm \sqrt{5})}{2}$ Factor.

$x = \dfrac{\cancel{2}(1 \pm \sqrt{5})}{\cancel{2}} = 1 \pm \sqrt{5}$ Simplify.

Your turn:

6. Solve $5z^2 - 2z = \frac{1}{5}$ using the quadratic formula.

	Answer	Text Ref	Video Ref		Answer	Text Ref	Video Ref
1	$\dfrac{-1 \pm \sqrt{37}}{6}$	Ex 1, p. 1191		**4**	$-2 \pm \sqrt{7}$		Sec 16.3, 2/4
2	$\dfrac{-7 \pm \sqrt{37}}{6}$		Sec 16.3, 1/4	**5**	$1 \pm \sqrt{5}$	Ex 5, p. 1193	
3	$-\dfrac{1}{2}; 5$	Ex 2, p. 1191–1192		**6**	$\dfrac{1 \pm \sqrt{2}}{5}$		Sec 16.3, 3/4

☐   **Next, insert your homework.** Make sure you attempt all exercises asked of you and show all work, as in the exercises above. Check your answers if possible. Clearly mark any exercises you were unable to correctly complete so that you may ask questions later. DO NOT ERASE YOUR INCORRECT WORK. THIS IS HOW WE UNDERSTAND AND EXPLAIN TO YOU YOUR ERRORS.

Section 16.4 Graphing Quadratic Equations in Two Variables

Before Class:

☐   Read the objectives on page 1200.

☐   Read the **Helpful Hint** boxes on pages 1201, 1202 and 1205.

☐   Complete the exercises:

Simplify the expression $x = \dfrac{-b}{2a}$ for the values of a and b.

   1. $a = 8,\ b = 2$ 2. $a = -8,\ b = -2$ 3. $a = -8,\ b = 2$

During Class:

☐   **Write your class notes.** Neatly write down **all** examples shown as well as key terms or phrases with definitions. If not applicable or if you were absent, watch the Lecture Series (DVD/CD) for this section and do the same (write down the examples shown as well as key terms or phrases). Insert more paper as needed.

Class Notes/Examples	Your Notes

Answers: **1)** $x = -2$ **2)** $x = -2$ **3)** $x = 2$

433

Section 16.4 Graphing Quadratic Equations in Two Variables

Class Notes (continued)	**Your Notes**

(Insert additional paper as needed.)

Section 16.4 Graphing Quadratic Equations in Two Variables

Practice:

☐   Next, complete any incomplete exercises below. Check and correct your work using the answers and references at the end of this section.

Review this example:

1. Graph the quadratic equation by finding and plotting ordered pair solutions.

$y = x^2 - 4$ Begin by finding the intercepts.

To find the y-intercept, we let $x = 0$.

$y = 0^2 - 4 = -4$

The y-intercept is $(0, -4)$.

To find the x-intercept, we let $y = 0$.

$0 = x^2 - 4 = (x + 2)(x - 2)$

$x - 2 = 0$ or $x + 2 = 0$

$\quad x = 2$ or $\quad\quad x = -2$

There are two x-intercepts: $(-2, 0)$ and $(2, 0)$.

Selecting additional values of x, and finding the corresponding y-values gives a table of points on the graph. The graph is called a parabola. The vertex of the parabola is $(0, -4)$.

Your turn:

2. Graph $y = 2x^2$ by finding and plotting ordered pair solutions.

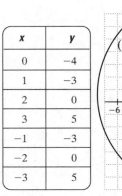

x	y
0	−4
1	−3
2	0
3	5
−1	−3
−2	0
−3	5

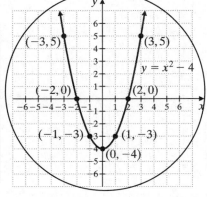

435

Section 16.4 Graphing Quadratic Equations in Two Variables

Complete this example:

3. Graph: $y = x^2 - 6x + 8$

Vertex formula: $x = \dfrac{-b}{2a} = \dfrac{-(-6)}{2 \cdot 1} = 3$ b = -6

Find the y-value of the vertex.

$y = x^2 - 6x + 8 = (3)^2 - 6(3) + 8$ Replace x with 3.

$y = 9 - 18 + 8 = -1$

The vertex is $(3, -1)$.

x	y
3	−1
4	0
2	0
0	8
1	3
5	3

y-intercept: $(0, 8)$

x-intercepts: $(2, 0), (4, 0)$

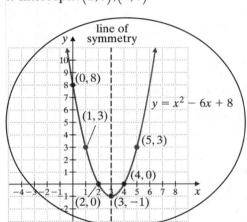

Your turn:

4. Graph: $y = -x^2 + 6x - 8$

Determine the vertex and the intercepts.

	Answer	**Text Ref**	**Video Ref**		**Answer**	**Text Ref**	**Video Ref**
1	See table and graph on previous page.	Ex 2, p. 1201		**3**	See table and graph above.	Ex 3, p. 1203	
2	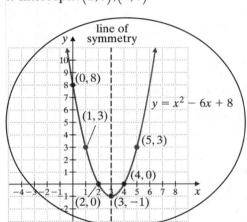		Sec 16.4, 1/3	**4**	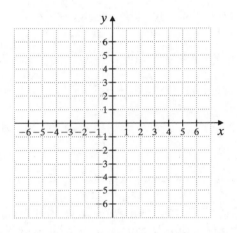		Sec 16.4, 2/3

☐   **Next, insert your homework.** Make sure you attempt all exercises asked of you and show all work, as in the exercises above. Check your answers if possible. Clearly mark any exercises you were unable to correctly complete so that you may ask questions later. DO NOT ERASE YOUR INCORRECT WORK. THIS IS HOW WE UNDERSTAND AND EXPLAIN TO YOU YOUR ERRORS.

436

Section 16.5 Interval Notation, Finding Domain and Ranges from Graphs, and Graphing
Piecewise-Defined Functions

Before Class:

☐   Read the objectives on page 1208.

☐   Read the **Helpful Hint** box on pages 1209.

☐   Complete the exercises:

1. Which inequality corresponds to the given graph?

a. $\{x \mid x < 2\}$ b. $\{x \mid x \le 2\}$ c. $\{x \mid x > 2\}$ d. $\{x \mid x \ge 2\}$

2. Which inequality corresponds to the given graph?

a. $\{x \mid x < 2\}$ b. $\{x \mid x \le 2\}$ c. $\{x \mid x > 2\}$ d. $\{x \mid x \ge 2\}$

During Class:

☐   **Write your class notes.** Neatly write down **all** examples shown as well as key terms or
phrases with definitions. If not applicable or if you were absent, watch the Lecture Series
(DVD/CD) for this section and do the same (write down the examples shown as well as key
terms or phrases). Insert more paper as needed.

Class Notes/Examples	**Your Notes**

Answers: **1)**   *d*   **2)**   *a*

Section 16.5 Interval Notation, Finding Domain and Ranges from Graphs, and Graphing Piecewise-Defined Functions

Class Notes (continued)	**Your Notes**

(Insert additional paper as needed.)

Section 16.5 Interval Notation, Finding Domain and Ranges from Graphs, and Graphing Piecewise-Defined Functions

---

Practice:

☐   Next, complete any incomplete exercises below. Check and correct your work using the answers and references at the end of this section.

Review this example:
1. Graph the solution set of the inequality on a number line and then write it in interval notation.

a. $\{x \mid x < -1\}$

b. $\{x \mid 0.5 < x \le 3\}$

The solution set of the inequality consists of all numbers less than -1, but not including -1. Instead of an open circle at -1, we use a parenthesis.

The solution set of the inequality consists of all numbers greater than 0.5 and less than or equal to 3. Instead of an open circle at 0.5, we use a parenthesis, and instead of a closed circle at 3, we use a bracket.

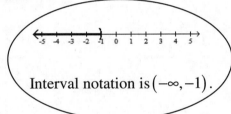

Interval notation is $(-\infty, -1)$.

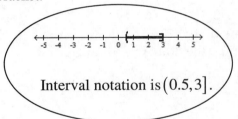

Interval notation is $(0.5, 3]$.

Your turn:
2. Graph the solution set of the inequality on a number line and then write it in interval notation.

a. $\{x \mid x < -3\}$

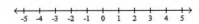

Interval notation is _____.

Section 16.5 Interval Notation, Finding Domain and Ranges from Graphs, and Graphing
Piecewise-Defined Functions

Review this example:

3. Find the domain and range.   The solution is below:

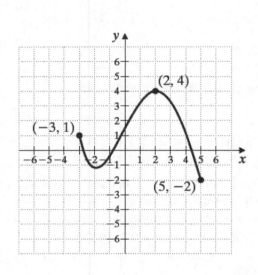

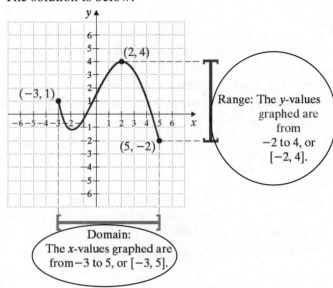

Range: The *y*-values graphed are from −2 to 4, or [−2, 4].

Domain: The *x*-values graphed are from −3 to 5, or [−3, 5].

Your turn:

4. Find the domain and range.

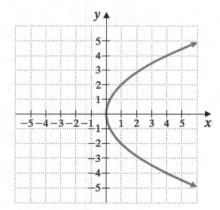

Section 16.5 Interval Notation, Finding Domain and Ranges from Graphs, and Graphing
Piecewise-Defined Functions

Complete this example:

5. Find the domain and range.   The solution is below:

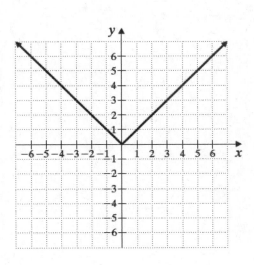   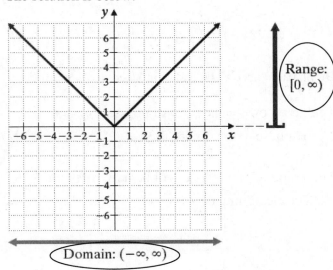

Your turn:

6. Find the domain and range.

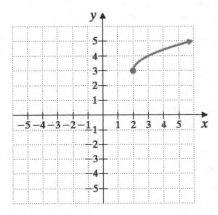

Section 16.5 Interval Notation, Finding Domain and Ranges from Graphs, and Graphing Piecewise-Defined Functions

Complete this example:

7. Graph the piecewise-defined function.

$$f(x) = \begin{cases} 2x+3 & if \ x \le 0 \\ -x-1 & if \ x > 0 \end{cases}$$

Graph each piece. The end points occur when $x = 0$.

The graph of $y = 2x + 3$ will be a ray with a closed circle at the point $(0, 3)$.

The graph of $y = -x - 1$ will be a ray with an open circle at the point $(0, -1)$.

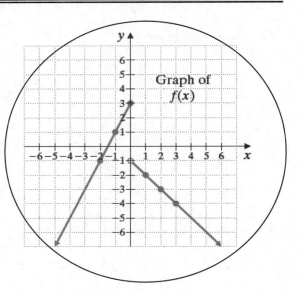

Graph of $f(x)$

If $x \le 0$,

$$f(x) = 2x + 3$$

Values ≤ 0

x	f(x) = 2x + 3
0	3 Closed circle
−1	1
−2	−1

If $x > 0$,

$$f(x) = -x - 1$$

Values > 0

x	f(x) = −x − 1
1	−2
2	−3
3	−4

Your turn:

8. Graph the piecewise-defined function.

$$f(x) = \begin{cases} x+3 & if \ x < -1 \\ -2x+4 & if \ x \ge -1 \end{cases}$$

Graph each piece. The end points occur when $x = -1$.

442

Section 16.5 Interval Notation, Finding Domain and Ranges from Graphs, and Graphing Piecewise-Defined Functions

	Answer	Text Ref	Video Ref		Answer	Text Ref	Video Ref
1	$(-\infty, -1)$ $(0.5, 3]$	Ex 2, 3, p. 1209		5	Domain: $(-\infty, \infty)$ Range: $[0, \infty)$	Ex 5, p. 1210	
2	$\{x \mid x < -3\}$ $(-\infty, -3)$		Sec 16.5, 1/7	6	Domain: $[2, \infty)$ Range: $[3, \infty)$		Sec 16.5, 6/7
3	Domain: $[-3, 5]$ Range: $[-2, 4]$	Ex 4, p. 1210		7	Graph of $f(x)$	Ex 9, p. 1211	
4	Domain: $[0, \infty)$ Range: $(-\infty, \infty)$		Sec 16.5, 3/7	8			Sec 16.5, 7/7

☐ **Next, insert your homework.** Make sure you attempt all exercises asked of you and show all work, as in the exercises above. Check your answers if possible. Clearly mark any exercises you were unable to correctly complete so that you may ask questions later. DO NOT ERASE YOUR INCORRECT WORK. THIS IS HOW WE UNDERSTAND AND EXPLAIN TO YOU YOUR ERRORS.

Preparing for the Chapter 16 Test

Start preparing for your Chapter 16 Test as soon as possible. Pay careful attention to any instructor discussion about this test, especially discussion on what sections you will be responsible for, etc.

☐ Work the Chapter 16 Vocabulary Check on page 1217.

☐ Read both columns (Definitions and Concepts, and Examples) of the Chapter 16 Highlights starting on page 1217.

☐ Read your Class Notes/Examples for each section covered on your Chapter 16 Test. Look for any unresolved questions you may have.

☐ Complete as many of the Chapter 16 Review exercises as possible (page 1219). Remember, the odd answers are in the back of your text.

☐ **Most important:** Place yourself in "test" conditions (see below) and work the Chapter 16 Test (pages 1223 - 1225) as a practice test the day before your actual test. To honestly assess how you are doing, try the following:
   - Work on a few blank sheets of paper.
   - Give yourself the same amount of time you will be given for your actual test.
   - Complete this Chapter 16 Practice Test without using your notes or your text.
   - If you have any time left after completing this practice test, check your work and try to find any errors on your own.
   - Once done, use the back of your book to check ALL answers.
   - Try to correct any errors on your own.
   - Use the Chapter Test Prep Video (CTPV) to correct any errors you were unable to correct on your own. You can find these videos in the Interactive DVD Lecture Series, in MyMathLab, and on YouTube. Search Martin-Gay Prealgebra & Introductory Algebra and click "Channels."

I wish you the best of luck....Elayn Martin-Gay